SpringerBriefs in Cybersecurity

T0171965

Cybersecurity is a difficult and complex field. The technical, political and legal questions surrounding it are complicated, often stretching a spectrum of diverse technologies, varying legal bodies, different political ideas and responsibilities. Cybersecurity is intrinsically interdisciplinary, and most activities in one field immediately affect the others. Technologies and techniques, strategies and tactics, motives and ideologies, rules and laws, institutions and industries, power and money—all of these topics have a role to play in cybersecurity, and all of these are tightly interwoven.

The *SpringerBriefs in Cybersecurity* series is comprised of two types of briefs: topic- and country-specific briefs. Topic-specific briefs strive to provide a comprehensive coverage of the whole range of topics surrounding cybersecurity, combining whenever possible legal, ethical, social, political and technical issues. Authors with diverse backgrounds explain their motivation, their mindset, and their approach to the topic, to illuminate its theoretical foundations, the practical nuts and bolts and its past, present and future. Country-specific briefs cover national perceptions and strategies, with officials and national authorities explaining the background, the leading thoughts and interests behind the official statements, to foster a more informed international dialogue.

More information about this series at http://www.springer.com/series/10634

Samer Al-khateeb • Nitin Agarwal

Deviance in Social Media and Social Cyber Forensics

Uncovering Hidden Relations Using Open Source Information (OSINF)

 Springer

Samer Al-khateeb
Department of Journalism,
Media & Computing
Creighton University
Omaha, NE, USA

Nitin Agarwal
Information Science Department
University of Arkansas at Little Rock
Little Rock, AR, USA

ISSN 2193-973X   ISSN 2193-9748   (electronic)
SpringerBriefs in Cybersecurity
ISBN 978-3-030-13689-5   ISBN 978-3-030-13690-1   (eBook)
https://doi.org/10.1007/978-3-030-13690-1

Library of Congress Control Number: 2019935524

This Springer imprint is published by the registered company Springer Nature Switzerland AG.
The registered company address is: Gewerbestrasse 11, 6330 Cham, Switzerland

Samer would like to thank his parents, Ihsan Al-khateeb and Basima Al-Qaraghuli, and his siblings, Samar, Saja, and Hasanain Al-khateeb, for all the care and support provided to him throughout his life. He also would like to especially thank his wife Lina Al Azzawi for all the love, patience, and positive vibes given to him at the time of writing this manuscript. Nitin Agarwal would like to dedicate this book to his family and friends with much love and gratitude for everything. Without all their encouragements, this endeavor would not be possible.

Foreword

This book describes the methodologies and tools used to collect, analyze, and visualize social media data and conduct social cyber forensic analysis. By applying these methodologies and tools on various events observed in the case studies contained within, their effectiveness is highlighted. The methodologies blend computational social network analysis and cyber forensic concepts and tools in order to identify and study information competitors. Through cyber forensic analysis, metadata associated with propaganda-riddled websites is extracted. This metadata assists in extracting social network information such as friends and followers; communication network information, i.e., networks depicting flows of information among the actors such as tweets, replies, retweets, mentions, and hyperlinks; and social media footprint, i.e., users' or groups' activity on other social media sites. Through computational social network analysis, the authors identify influential actors and powerful groups coordinating the disinformation campaign. A blended social cyber forensic approach allows them to study cross-media affiliations of the information competitors. For instance, narratives are framed on blogs and YouTube videos, and then Twitter and Reddit, for instance, will be used to disseminate the message. Social cyber forensic methodologies enable researchers to study the role of modern information and communications technologies (ICTs) in the evolution of information campaign and coordination. In addition to the concepts and methodologies pertaining to social cyber forensics, this book also offers a collection of resources for readers including several datasets that were collected during case studies, up-to-date reference and literature surveys in the domain, and a suite of tools that students, researchers, and practitioners alike can utilize. Most importantly, the book demands a dialogue between information science researchers, public affairs officers, and policymakers to prepare our society to deal with the lawless "wild west" of modern social information systems triggering debates and studies on cyber diplomacy.

Springer Nature, Cham, Switzerland   Christopher T. Coughlin
March 2018

Preface

In order to address the problem of deviance in social media, we need to understand the power of social media analytics, the importance of open-source information (i.e., information from publicly available sources), and the impact of combining both to study deviant groups and tactics. We wrote this book to raise awareness of the deviant usage of social media by various groups, both virtual or physical. Hence, we introduce and blend a set of graph-theoretic concepts, social network analysis, and social cyber forensic tools and methodologies.

The readers will understand the aforementioned concepts, analysis, and tools to collect various social media data, and the metadata associated with various entities then analyze the collected data and derive insights. The methodologies and tools that we introduce throughout the book have been tested on many real-world events, e.g., the anti-NATO propaganda campaigns, ISIS/Daesh/ISILs propaganda, and the cyber campaigns of Blackhat hackers on social media. These methodologies have been developed with underpinnings in sociological theories of collective action, cyber forensic science, graph theory, and social network analysis. At the end of this text, we present a set of case studies that bring together all the concepts introduced in this book.

This book is suitable for undergraduate and graduate students as well as analysts who are interested in studying the problem of deviance in social media. After reading this book, readers will be enriched with a knowledge of various off-the-shelf tools. We hope this book can jump-start their skills on collecting, analyzing, and visualizing various social media datasets and use social cyber forensic analysis to unveil hidden relations. The book is written considering the wide range of disciplines that would benefit from it. For this reason, the concepts and tools covered in this book require absolutely zero technical background.

Omaha, NE, USA   Samer Al-khateeb
Little Rock, AR, USA   Nitin Agarwal
December 2018

Acknowledgments

We thank many colleagues who made substantial contributions in various ways to this project. The members at the Collaboratorium for Social Media and Online Behavioral Studies (COSMOS) group at UA-Little Rock made this project much easier and enjoyable. We are grateful for their comments.

We really appreciate Springer and particularly Christopher Coughlin, Editor, for helping us throughout this project.

The research presented in this book is funded in part by the US National Science Foundation (IIS-1636933, IIS-1110868), US Office of Naval Research (N00014-10-1-0091, N00014-14-1-0489, N00014-15-P-1187, N00014-16- 1-2016, N00014-16-1-2412, N00014-17-1-2605, N00014-17-1-2675), US Air Force Research Laboratory, US Army Research Office (W911NF-16-1-0189), US Defense Advanced Research Projects Agency (W31P4Q-17-C-0059), Arkansas Research Alliance, the Jerry L. Maulden-Entergy Fund at the University of Arkansas at Little Rock, and Creighton University's College of Arts and Sciences. Any opinions, findings, and conclusions or recommendations expressed in this material are those of the authors and do not necessarily reflect the views of the funding organizations. The researchers gratefully acknowledge the support.

Last, and certainly not least, we thank our families, for supporting us through this fun but time-consuming project. We dedicate this book to them, with love.

Contents

About the Authors

Samer Al-khateeb is an Assistant Professor in the Department of Journalism, Media and Computing, College of Arts and Sciences, at Creighton University and a former Postdoctorate Research Fellow at the Collaboratorium for Social Media and Online Behavioral Studies (COSMOS) at the University of Arkansas at Little Rock (UA-Little Rock). He obtained his Ph.D. in Computer and Information Sciences, a master's degree in Applied Science, and a bachelor's degree in Computer Science from UA-Little Rock. He studies deviant acts (e.g., deviant cyber flash mobs and cyber propaganda campaigns) on social media that are conducted by deviant groups (e.g., Daesh, Black hat hackers, and Propagandist) which aim to influence individual's behaviors and provoke hysteria among citizens. He also studies the type of actors these deviant groups use to perform their acts, i.e., are they human (e.g., Internet trolls) or automated actors (e.g., social bots) by leveraging social science theories (e.g., the theory of collective action), social network analysis (e.g., centralities and community detection algorithms), and social cyber forensics (e.g., metadata collection to uncover the hidden relations among these actors across platforms). He has many publications including book chapters, journal papers (e.g., *Defence Strategic Communications*; *Journal of Digital Forensics, Security and Law*; *Journal on Baltic Security*; and the *IARIA (International Journal on Advances in Internet Technology)*), conference proceedings, and conference presentations. He won various awards such as the Staff Achievement Award for Educational Achievements, Excellence in Research Award, Outstanding Graduating Student Award (Master's Level), Who's Who Among Students in American Universities and Colleges, the Best Paper Award, 2nd Place Most Innovative Award, and 2nd Place Societal Impact Award, among others.

Nitin Agarwal is the Jerry L. Maulden-Entergy Endowed Chair and Distinguished Professor of Information Science at the University of Arkansas at Little Rock. He is the Founding Director of the Collaboratorium for Social Media and Online Behavioral Studies (COSMOS) at UA-Little Rock. His research aims to push the boundaries of our understanding of cyber social behaviors that emerge and evolve constantly in the modern information and communication platforms with applications in defense and security, health, business and marketing, finance, and education. At COSMOS, he is leading projects funded by over $10 million from an array of federal agencies, including US National Science Foundation, Office of Naval Research, Army Research Office, Air Force Research Laboratory, Defense Advanced Research Projects Agency, and Department of State, and plays a significant role in the long-term partnership between UA-Little Rock and the Department of Homeland Security. He developed publicly available social media mining tools, viz., Blogtrackers, YouTube Tracker, and Focal Structure Analysis used by NATO Strategic Communications and public affairs, among others. Dr. Agarwal participates in the national Tech Innovation Hub launched by the US Department of State to defeat foreign-based propaganda.

His research contributions lie at the intersection of social computing, behavior-cultural modeling, collective action, social cyber forensics, AI, data mining, and machine learning. From Saudi Arabian women's right to drive cyber campaigns to autism awareness campaigns to ISIS and anti-West/anti-NATO disinformation campaigns, at COSMOS, he is directing several projects that have made foundational and applicational contributions to social and computational sciences. He has published 8 books and over 150 articles in top-tier peer-reviewed forums with several best paper awards and nominations. Dr. Agarwal obtained Ph.D. from Arizona State University with outstanding dissertation recognition in 2009. He was recognized as one of "The New Influentials: 20 In Their 20s" by Arkansas Business in 2012. He was recognized with the university-wide Faculty Excellence Award in Research and Creative Endeavors by UALR in 2015. Dr. Agarwal received the Social Media Educator of the Year Award at the 21st International Education and Technology Conference in 2015. In 2017, the *Arkansas Times* featured Dr. Agarwal in their special issue on "Visionary Arkansans: A Celebration of Arkansans with ideas and achievements of transformative power." Dr. Agarwal was nominated as International Academy, Research and Industry Association (IARIA) Fellow in 2017, Arkansas Academy of Computing (AAoC) Fellow in 2018, and Arkansas Research Alliance (ARA) Fellow in 2018.

Visit http://ualr.edu/nxagarwal/ for more details.

Acronyms

API	Application Programming Interface
ASAs	Automated social actors/agents
BRJP	Brilliant jump exercise
CA	Collective Action
CCA	Cyber collective action
CFM	Cyber flash mob
DCFM	Deviant Cyber Flash Mob
DDoS	Distributed denial-of-service
DHN	Deviant hacker networks
EEAS	European External Action Service
FM	Flash mob
FSA	Focal Structures Analysis
ICSR	International Centre for the Study of Radicalization and Political Violence
ICT	Information and communications technology
I2P	Invisible Internet Project
IRC	Internet Relay Chat
ISIL	Islamic State of Iraq and the Levant
ISIS	Islamic State of Iraq and Syria
ITAR	Information Telegraph Agency of Russia
LIWC	Linguistic Inquiry and Word Count
MMOGs	Massive Multiplayer Online Games
MUDs	Multi-User-Domains
NATO	The North Atlantic Treaty Organization
NIGMS	US National Institute of General Medical Sciences
NodeXL	Network Overview, Discovery and Exploration for Excel
NRNB	National Resource for Network Biology
NSMT	New Social Movement Theory
ODBC	Open Database Connectivity
ODGs	Online Deviant Groups
ORA	Organizational Risk Analyzer

OSINF	Open-Source Information
OSN	Online Social Network
PAO	Public Affairs Officer
SCF	Social cyber forensics
SNA	Social network analysis
TAGS	Twitter Archiving Google Sheet
TASS	Telegraph Agency of the Soviet Union
TCOs	Transnational Crime Organizations
TOR	The Onion Router
TRJE	Trident Juncture Exercise
WCE	Web Content Extractor
WoW	World of Warcraft
WTC	Web Tracker Code

List of Figures

Chapter 1
Deviance in Social Media

Abstract In this chapter, we explain what we mean by deviance in social media. We give examples of four types of deviance as observed on social media, viz., *deviant acts*, *deviant events*, *deviant tactics*, and *deviant groups*. We provide historical information, definitions, and examples that will be studied/explained in more details throughout the book. This chapter would help the readers understand the scope of the problem of deviance in social media, familiarize with definitions and examples of deviant events, groups, acts, and tactics, and peek at the social science theories that can explain such emergent deviant behaviors on social media.

Keywords Deviance in social media · Online deviant behaviors · Deviant groups · Online deviant events · Online deviant tactics · Theory of collective action

1.1 Introduction

The Internet is indisputably one of the greatest inventions of the twentieth century that has revolutionized communication and information dissemination. Its affordability and ease of use nature have made tremendous contributions to human life in various aspects, such as education, health care, and business, among others. People nowadays can attend colleges online, track how many calories they consumed or burned using various mobile apps, and business owners can keep track of their stocks and the progress of their company by using the Internet. However, with all these benefits that come from using Internet technologies, there are also some adverse effects.

In the last decade and a half, a new form of information communication medium has emerged, i.e., online social networks (OSNs). This technology was initially developed with benign goals in mind, e.g., to socialize with friends and family members virtually; share pictures and videos; or connect with professionals to find a job; but nowadays it has become a fertile ground for malicious activities such as cyber-bullying, cyber-crime, cyber-terrorism, and cyber-warfare.

A rough estimate suggests that 2.5 million terabytes of data is created every day in 2017, with no signs of slowing down, in fact 90% of the data in the world was generated in the last 2 years. In addition to that, there are 3.7 billion humans connected to the Internet globally. These humans are generating a vast amount of data in the digital sphere using various applications such as IoT, scientific experiments data, emails, social media, among others. Social media plays an important role in generating a lot of data, for example, on a daily basis there are six hundred fifty-six million tweets sent, sixty-seven million Instagram photos are uploaded, and more than five billion videos are watched on YouTube in 2017 [1, 2]. This widespread use of contemporary forms of information and communications technology (ICTs), e.g., social media, has transformed the way people interact, communicate, and share information. It has afforded a fundamental paradigm shift in the coordination abilities of people leading to manifestations of cyber collective actions (CCA) [3, 4] in various forms (see Fig. 1.1). These can be *social movements*[1] for socio-political transformation; *campaigns* for better governance through citizen journalism and engagement; *Flash Mobs (FM)* for promoting a cause for entertainment, or for deviant purpose; *Parkour*[2] as sportive practice; or *deviant collective hacking* activity to steal personal information or leak classified documents (also known as hacktivism).

The problem of deviance in social media is broad and big. Many researchers have studied different parts/aspects of deviance. Many researchers attempted to come up with countermeasures that would help mitigate the risks posed by this problem. Some of the attempts to stop deviance in social media or at least to raise awareness of the problem are: StopFake.org (an online crowdsourcing-based efforts), the European External Action Service (EEAS) [8], the East Strategic Communication Task Force and its program to fight disinformation (EUvsDisInfo) [9], and the Estonian organization PropaStop.Org which was created to identify and debunk fake imagery and stories about the war in Ukraine. However, such efforts are severely

---

[1] A very well-known example of social movements that succeeded due to the role of social media— as it acted as a vehicle for change in many countries—was during the so-called the "Arab Spring." When Mr. Mohamed Bouazizi, a 26-year-old Tunisian fruit vendor set himself on fire, on December 18, 2010, in a protest in front of a government building. This led to many protests and grievance that caused President Zine El Abidine Ben Ali to step down. That act was captured on camera and disseminated on social media, which encouraged many other countries to protest against the authoritarian regimes in the Middle-East [5].

[2] Parkour is the manifestation in which *traceurs* (parkour practitioners) jump between distance rooftops, climb vertical walls, and continuously search for new ways to challenge the rules of gravity. Parkour history is more than 100 years as Michael Atkinson states in his article "Parkour, Anarcho-Environmentalism, and Poiesis" [6]. This practice goes back to a style called *Hebertism* that emerged at the beginning of the twentieth century. The practice contains variety of obstacles and landscapes usually in the wooded setting "as an unfettered animal" [7]. The first parkour was formed originally in Paris and since then it emerged all over the world, from Canada and USA to Russia and the Philippines.

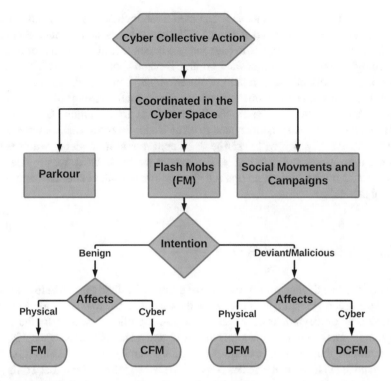

Fig. 1.1 The different forms of cyber collective action (CCA), e.g., parkour, social movements and campaigns, and flash mobs (FM) by their types, i.e., benign flash mobs (FM), cyber flash mobs (CFM), and deviant cyber flash mobs (DCFMs)

limited and easily outnumbered by various online deviant groups that are aided by vast armies of social bots and Internet trolls. Next, we provide a literature review of the problem of deviance in social media and explain what we mean by online deviant groups (ODGs), online deviant events/acts, and online deviant tactics. Then we take a little detour into the social science theories that can help understand deviance in social media.

1.2 Literature on Online Deviant Behaviors

According to Oxford Dictionaries the word "deviance" means the "state of diverging from the normal, usual, or accepted standards" in any social space, e.g., a society [10]. This deviance sometimes leads to a good or beneficial outcome to the society such as innovation or creativity but in many cases, it leads to harmful effects on the society as in the case of crimes [11]. Throughout this book, we use the word "deviance" to refer to the latter case, i.e., the unusual, unaccepted, illegal, or harmful effects on the society. Deviance in social media can include deviant groups, deviant acts, deviant events, or deviant tactics. The deviant acts and deviant events are usually carried on by groups who have interest in conducting such acts for financial or ideological purposes, i.e., deviant groups. Deviant groups are usually connected through some sort of social relations, for example, family or tribal membership, friendship, colleagues at work, classmates, or virtually through different social media platforms, such as Facebook, Twitter, Instagram, Tumblr, Snapchat, etc. These deviant groups sometimes use deviant tactics to make their deviant events or acts successfully. We will describe these concepts in more depth with the help of several examples and case studies throughout the book. Next, we provide more details about online deviant groups (ODGs), online deviant events/acts, and online deviant tactics.

1.2.1 Online Deviant Groups

We define online deviant groups (ODGs) as groups of individuals who are connected online using social media platforms or the dark web[3] and have interest in conducting deviant acts or deviant events that can cause significant danger to the public in general. These ODGs could include state and non-state actors (see Fig. 1.2). For example, the so-called Islamic State in Iraq and Levant (ISIL), anti-NATO propagandist [17], Deviant Hackers Networks (DHNs) [18], and Internet Trolls [19]. Next, we provide more information about the aforementioned ODGs.

1.2.1.1 Islamic State of Iraq and the Levant/Syria (ISIL) (ISIS) (Daesh)

There are many examples of ODGs. One of the most heinous groups is Daesh. Since the fall of Mosul—the second largest city in Northern Iraq [20]—at the hands of Daesh the Iraqi army[4] was struggling in the fight against the extremist group and in many cases was losing battles, e.g., the fall of the city of Ramadi in the west of Iraq [21]. Several journalistic accounts have tried to identify the reasons for the

---

[3]The dark web constitutes only one part of what is called the invisible web[12], hidden web [13], or deep web[14]. The dark web websites use anonymity tools such as the "packet based routing," i.e., the Invisible Internet Project (I2P) and the "circuit based routing," i.e., The Onion Router (TOR) to conceal their IP address [15]. The dark web is known for having markets that sell various illegal products such as drugs, guns, and even child pornography. In addition to having these types of markets, many people use the dark web because it provides protection against censorship or surveillance [15]. The invisible, hidden, or deep web is a part of the World Wide Web that is not indexed by any standard search engine. The "surface web" is the opposite of the "deep web," i.e., the part of the World Wide Web that is indexed by standard search engines [16].

[4]Iraq Prime Minister Declares Victory Over ISIS. The NY Times, December 9, 2017. Available at https://goo.gl/AkUJkT.

Fig. 1.2 Some examples of online deviant groups (ODGs)

Iraqi army to flee, the financial sources used by ISIL to cover operational costs [22], and their power as a group to recruit new fighters who help them fill the place of the ones they lose in the battlefield [23]. Additionally, many reports have highlighted the sophisticated use of social media by ISIL and their ability to spread their message in large scale during a short period of time [24] or to recruit new members using social media.

The process of organized propaganda dissemination by a group of individuals using social media, e.g., Twitter, is an instance of a deviant cyber flash mob (DCFM) (the reasons for this assertion are explained in Sect. 1.2.2) [25]. For example, the dissemination of ISIL's beheading video-based propaganda of the Egyptians Copts in Libya [26], the Arab-Israeli "Spy" in Syria [27], and the Ethiopian Christians in Libya [28]. ISIL's Internet recruitment propaganda or the E-Jihad is very effective in attracting new group members [29]. For example, a study conducted by Quiggle [30] on the effects of developing high production value beheading videos and releasing on social media by ISIL members shows that ISIL's disseminators are excellent narrators and they choose their symbols very carefully to give the members of the groups the feeling of pride as well as cohesion. The beheading of civilians has been studied in the literature by Regina Jones [31]. In her study, Jones categorized the reasons for why beheading is done into four main categories, viz., *Judicial, Sacrificial, Presentational*, and *Trophy*. ISIL's communicators designed the beheading videos to serve all of the four categories. First, it is considered as *judicial* because in some countries that are ruled by *Sharia law* criminals who are found guilty of burglary, apostasy, drug trafficking, witchcraft, murder, and rape are beheaded. Second, by putting the video on YouTube gives ISIL's remote members a source of voyeuristic pleasure, i.e., *presentational* category. Third, beheading is considered by ISIL's ideology as a communal blood ritual that cleanses European-born ISIL's members of "Westoxification,[5]" i.e., *sacrificial* category. Finally, it is considered as a *trophy* because it increases the prestige of ISIL's members, who beheaded the hostages among their fellow jihadists.

1.2.1.2 Anti-NATO Propagandist

In addition to ISIL's sophisticated use of social media, online propagandists are also using social media to manipulate or influence people's opinions about specific events by disseminating propaganda, misinformation, and disinformation about these events. One example of the online propagandists group is anti-NATO propagandist. Anti-NATO propagandists are groups of individuals who use social media in a sophisticated way to disseminate disinformation to a large global audience about different NATO activities such as their military exercises in Europe. These groups usually try to project an image that NATO exercises in Europe are a preparation for a Third World War (WWIII) or trying to provoke Russia.

1.2.1.3 Deviant Hackers Networks (DHNs)

Blackhat hackers are also considered as ODGs. Blackhat hackers are groups of hackers who coordinate cyberattacks with malicious intent. Hacker groups use dark web and social media to communicate and coordinate their acts. The threat these Blackhat hackers pose is real and can manifest in many ways, such as disabling critical infrastructure. In one incident the Ukrainian citizens suffered power outage during Christmas eve because of a Russian-sponsored hacker group that coordinated a cyberattack on the power grids in December 2015.[6] In addition to coordinating cyberattacks [18], Blackhat hackers also use social media to sell various programs or software that can capture sensitive financial data [32], sell the financial data on online forums or dark markets to make a profit [33], or organize events where they ask their supporters to do something deviant in nature.

One of the Blackhat hacker groups that use social media to organize events is the Anonymous. They ask their supporters to include a specific hashtag, e.g.,

---

[5]Westoxification is the state of being inebriated with Western culture and ideas.

[6]U.S. helping Ukraine investigate power grid hack. *Reuters*. January 12, 2016. Available at http://reut.rs/1PqNAYG.

Fig. 1.3 The banner Anonymous designed for their operation

#OpIsrael[7] and #OpRohingya,[8] in their message or tweet and then post their message on a specific official Twitter or Facebook page at a specific time. They do this to raise awareness, show their solidarity, or antagonism for various issues, or to perform a distributed denial-of-service (DDoS) attack. For example, in one instance they asked their supporters to attack a Saudi Government website[9] as an expression of their support for a Saudi youth who was captured at the age of 17 participating in a protest against the Saudi Government and sentenced to death. Figure 1.3 shows the banner designed by Anonymous for their #OpNimr campaign on Twitter to support the Saudi youth against the Saudi Government. Note that they associated their hashtag, i.e., *#OpNimr*, with one of the trending hashtags in 2016, i.e., *#RIO2016*, to spread their message to all the people who are interested in the 2016 Rio Olympic Games. This is a very well-known tactic to make a message reach as many people as possible and is known as *hashtag latching*. Sometimes Anonymous also use keywords in their tweets to communicate a message that no one would understand unless they are from the same group, e.g., the *hashtag* "Tango Down" is used by Anonymous hacking group to signal a successful DDoS attack[10] (Tango is a keyword that is used to eliminate a target. So Tango Down implies target down.). Anonymous also use specific hashtags to signify the beginning of their campaign, e.g., the hashtag "#OpBeast[11]" was used during the attack that was planned by Anonymous against animal cruelty and depravity websites. Anonymous were encouraging people to tweet using this hashtag to spread awareness that they

---

[7]Anonymous is hacking Israeli Web sites. *The Washington Post*. November 17, 2012. Available at https://www.washingtonpost.com/news/worldviews/wp/2012/11/17/anonymous-is-hacking-israeli-web-sites/.

[8]Anonymous uses Twitter to highlight humanitarian crisis in Burma. *The Verge*. March 26, 2013. Available at https://www.theverge.com/2013/3/26/4148908/anonymous-oprohingya-burma-myanmar-humanitarian-crisis-campaign.

[9]#OpNimr: Anonymous fight to stop execution of Saudi youth. Al-Jazeera, September 28, 2015. Available at http://stream.aljazeera.com/story/201509282137-0025017.

[10]Anonymous: CIA, Interpol websites "tango down." *Reuters*. 2012.

[11]Anonymous launched #OpBeast against animal cruelty and depravity. *TechWorm*. 2015.

are against some of the websites that do not operate in the dark web (some .com websites) which provide services that show images of animal cruelty, feature animal depravity, and bestiality [18].

1.2.1.4 Internet Trolls

Internet trolls are yet another type of online deviant groups (ODGs) who flourished as the Internet became more social, i.e., with the advent of social media. The word "troll" refers to a mythical creature from Scandinavian folklore who is slovenly, hideous, angry, live usually in dark places, and can eat anything to survive [34]. Internet trolls are called this way because they hide behind their computer screen, angry, and troublesome in many ways. Internet trolls are formally defined as *a person who disseminates provocative posts on social media for the troll's amusement or because (s)he was paid to do so* [34–36]. These individuals disseminate provocative posts, e.g., insulting a specific person or group, posting false information, or propaganda on popular social media sites, which results in a flood of angry responses and often hijack the discussion for their amusement or for financial incentives [19, 35, 36]. There are few examples where these individuals were paid to do such a disruptive behavior, e.g., the Russians are spending millions of dollars to finance the Kremlin's Troll Army (legions of pro-Russia and English-speaking Internet commenter's). Their goal is to promote president Vladimir Putin and his policies, and to spread disinformation about some events or disseminate propaganda war on Ukraine [19]. In the recent Ukraine–Russia conflict, sites like VKontakte (a Russian social media platform), LiveJournal, Twitter, YouTube, and Tumblr, etc., have been used as propaganda machines to justify the Kremlin's policies and actions [37]. According to Interpret Magazine, Kremlin recruited over 250 Internet Trolls to disseminate false information, rumors, or propaganda on popular blogs with large audiences and paid each of them $917 per month to work around the clock producing posts on social and mainstream media. These Internet trolls would create a stream of invective against pro-Ukrainian media and Western news sources. They wrote unflattering posts about Russia and posted numerous comments and blog posts each day using multiple "sock puppet" accounts that work in small groups, e.g., triads (a group of three individuals). Such "troll armies" (or "web brigades") piggyback on the popularity of social media to disseminate fake pictures and videos coordinating effective disinformation campaigns to which even legitimate news organizations sometimes fall prey [19].

1.2.2 Online Deviant Events

Since the occurrence of the first "flash mob" (FM) organized by Bill Wasik in Manhattan in 2003, flash mob phenomenon has become widespread. Recent journalistic accounts have reported that this form of public engagement has the potential to pose considerable amounts of risks to civil, political, social, and economic stability of a region. This raises the importance of systematically studying such behaviors. Modern information and communication technologies (ICTs) provide affordable and easy to use means of communications (such as social network platforms, viral emails, and SMS) that facilitates the process of recruiting, training, and looking for a specific sector of the society (specific gender, age, political affiliation, interest, and cultural background) easier than it was before. This in turn has led to an increase in the occurrences of emerging socio-technical behaviors [7], such as flash mobs (FM), cyber flash mobs (CFM), and deviant cyber flash mobs (DCFMs).

Flash mob (FM) is a form of public engagement, which according to Oxford Dictionaries is defined as "a large public gathering at which people perform an unusual or seemingly random act and then quickly disperse" [38]. Recent observations pertaining to the deviant aspect of the flash mobs has insisted to add a highly debated perspective, which is *the nature* of the flash mob whether it is for entertainment, satire, and artistic expression or it is a deviant act that can lead to robberies and thefts, also known as criminal flash mobs. A flash mob could be entertaining such as gathering and dancing in a shopping mall (see Fig. 1.4), or *happy birthday for a bus driver* [39] or it can be the new face of transnational crime organizations (TCOs) [40], such as the "Bash mob" (see Fig. 1.5) that happened in Long Beach, California, in July 9, 2013 [41].

ODGs in most cases organize events in a collective manner, i.e., they invite or recruit people who in some cases have close ties with—or sometimes random people—to participate in an organized event that has a specific goal, e.g., inviting people on social media to attack a specific Twitter or Facebook page[18]. These organized events are called *Cyber Flash Mobs* (CFMs) if they are coordinated using social media platforms and have *no harmful effect* on the society, but if they

Fig. 1.4   A dancing flash mob at Toronto Eaton Centre

LBPD Prepared For Potential Bash Mob Event

2013-07-19 · By Editor   ✉

share this: 👍 Like 2   🐦 Tweet   5   📌 Pinit   in Share   G+

Long Beach Police Department. Photo by Barbara Holbrook, Everything Long Beach

On July 9th at about 8:00 p.m., well over 100 individuals decided to participate in a "Bash Mob" in the Pike and City Place area of our downtown. A "Bash Mob" is a planned, sudden assembly of individuals who attack innocent people and businesses by committing thefts, property damage, and assaults. Once these crimes are committed, the individuals flee the area. "Bash Mobs" have been seen throughout the nation and are a cause for great concern. Many people do not realize the mere participation in such an event can result in felony charges including conspiracy, and are punishable by imprisonment in the state prison.

Fig. 1.5 Bash mob in Long Beach California as an example of a deviant flash mob

do have a goal to harm, damage, or conduct illegal activities, i.e., have a deviant intent, then we call it *Deviant Cyber Flash Mobs* (DCFMs). DCFMs are defined as the cyber manifestation of flash mobs (FM). They are known to be coordinated via social media, telecommunication devices, or emails and have a harmful effect on one or many entities such as government(s), organization(s), society(ies), and country(ies). These DCFMs can affect the physical space, cyberspace, or both, i.e., the "cybernetic space" [42].

The phenomenon of deviant cyber flash mob (DCFM) increased rapidly since its inception. As soon as mobile devices and texting became common in the early

2000s, people realized the potential of these technologies to mobilize individuals to act in a coordinated manner—possibly disruptively—and then quickly disperse into the anonymity of the Internet [43]. Theories such as collective action [44], collective identity formation [45, 46], collective decision [47], and social capital [47, 48] have often been used to explain the group dynamics underlying collective behavior phenomena. Collective action (CA) is defined as all activity of common or shared interest among two or more individuals [49]. Collective action theories can be traced back to Ronald Coase's [50] economic explanation on why individuals are willing to form companies and partnerships in business trading other than making bilateral contracts between individuals. Many years later, many collective action theories were developed such as the neutrality theorem [51, 52], club theory [53], folk theorem [54], and the most eminent one, rational choice theory [55]. As the collective action studies developed, many other approaches emerged such as the Resource Mobilization Theory [56–58] that emerged in the 1970s and early 1980s. This theory was the most influential approach in explaining the success of a collective action. However, this theory and the rational choice theory were criticized for failing to answer the question as to *"how social meaning is constructed and how it works as a driving force for action?"* As a reaction to this critique, the new social movement theory (NSMT) emerged. This theory associates actions with belief systems that revolve around a set of values and symbols that are specific to the group [59, 60]. These theories and concepts have been studied in the past by many researchers, but to the best of our knowledge never been used to study DCFMs. Hence, in Sect. 1.3 we leverage the theory of collective action to model DCFMs along with a CFM scenario. The complete details of this study are published in [61] and a successful operationalization of the model on real world data is published in [25]. Here, we briefly highlight the model. Interested readers are encouraged to read the aforementioned publications.

1.2.3   Online Deviant Tactics

Most of the aforementioned ODGs use various *tactics* to help them spread their message and recruit new people. One of the most used tactics is called *deviant social bots*. Bots are computer programs that can be designed and scheduled to perform various tasks on behalf of the bot's owner or creator. Research shows that most Internet traffic, especially on social media, is generated by bots [62]. Bots can have "**good**" intention such as *informational bots*; *chatbots*[12]; *entertainment bots* (game bots and art bots); *crawlers*; and *transactional bots*. However, bots could also have a "**bad**" or deviant intent such as *scraper bots*, *hacker bots*, *spambots*, and *impersonators bots* which we call *deviant social bots* [63]. Impersonators bots

---

[12]Woebot is a chatbot that helps people track their mood and give them therapeutic advices, available at https://woebot.io.

are computer software that are designed to act like a human in social spaces, e.g., social media. These impersonators bots if they mimic human and try to do deviant behavior on social media we call them as deviant social bots. Deviant social bots can influence people's opinion by disseminating propaganda or disinformation.

Bots or automated social actors/agents (ASAs) are not a new phenomenon. They have been studied previously in literature in a variety of domains. One of the earliest bots emerged in 1993 in the Internet Relay Chat (IRC), which was called Eggdrop [64]. This bot or ASA had very simple tasks which were basically welcoming and greeting the new participants and warn them about the other users' actions [64]. Shortly after that, the usage of bots in IRC became very popular due to the simplicity of implementing them [65]. Bots evolved over time, i.e., they have more functionalities and the tasks these bots were assigned became more complicated and sophisticated. Bots are also used in the multi-user-domains (MUDs) and massive multiplayer online games (MMOGs) [66]. The emergence of multi-user domains emphasized the need for automated social actors (ASAs) to enhance the playing experience. As the online gaming market grew, the need for more advanced bots increased. In MMOGs, such as the World of Warcraft (WoW) unauthorized game bots emerged. These unauthorized bots enhance and trigger mechanisms for the players by often sitting between the players' client application and the game server. These bots would play the game autonomously in the absence of the real player. In addition, some bots were also able to affect the game ecologies, i.e., amassing experience points or game currency (e.g., virtual gold) [67].

Social media has emerged within the last decade and a half and bots have been observed fairly recently. In a study conducted by Facebook in 2017, about 12% of all Facebook users are fake and duplicate accounts. This means that there are about 270 million user accounts on Facebook that do not belong to real people [68]. In addition to Facebook, another study conducted in the same year suggests that 15% of Twitter accounts are bots rather than real people, which means 48 million accounts are not human [69]. Some of these bots are very sophisticated and some even try to mimic human behavior, which makes any task to discover, detect, or capture them a challenging one [70]. Abokhodair et al. [67] studied the use of social bots during crises in the 2012 Syrian Civil War conflict. Their study focused on one bot that lived for 6 months before Twitter detected and suspended it. The study analyzed the life and the activities of that bot. More focus was given to the content of the tweets, i.e., they classified the content of the tweets into 12 categories, viz., news, opinion, spam or phishing, testimonial, conversation, breaking news, mobilization of resistance or support, mobilization for assistance, solicitation for information, information provisioning, pop culture, and other. Authors also classified bots based on the content posted, the time before the bot gets suspended, and type of activity the bot does (tweet or retweet) into the following categories:

- *Core Bots*: are Twitter accounts that form the core of the network and have three subcategories:

  - *Generator Bots*: tweet a lot but seldom retweet anything (like content generators).

- *Short-Lived Bots*: retweet a lot but seldom tweet and lasted for less than 6 weeks before Twitter suspended the account (like content amplifiers).
- *Long-Lived Bots*: retweet a lot but seldom tweet and lasted for more than 25 weeks before Twitter suspended the account (like content amplifiers).

- **Peripheral Bots**: are Twitter accounts being lured to participate in the dissemination process. These accounts look more like human Twitter accounts and they retweet one or more tweets generated by the core bots.

More sophisticated bots were used to disseminate propaganda during the 2014 Crimean Water Crisis, the NATO's Trident Juncture 2015 Exercise, the 2015 ISIS video-based beheading propaganda, and the NATO's Dragoon Ride 2015 Exercise. All the case studies are discussed in more details in Chap. 5.

1.3   Leveraging the Theory of Collective Action to Study DCFMs

In the book "The Mathematics of Collective Action" [44], Coleman proposed a framework for collective action and provided 25 mathematical definitions. Concepts such as *Power*, *Control*, *Utility*, and *Interest* defined by Coleman are borrowed here to derive the model that explains DCFM behavior. The concepts have been appropriately modified to suit the cyberspace, especially for the DCFM behavior. We revisit the theories and concepts that have been studied in social science and assess their applicability to the cyberspace. Before we introduce the model we first list the high-level relations between the aforementioned concepts. The model presented below for DCFM would work equally well for a CFM scenario. As a matter of fact, at the end of this section, we demonstrate how the model can be applied to assess the case of a CFM scenario via a toy example. The relations provided below help us assess the motivation needed to sustain coordinated acts such as DCFM and govern its success or failure:

1. A DCFM is more important when many actors are interested in participating. Also, more actors will be interested to participate in an important DCFM.
2. The interest of an actor in a DCFM increases as the utility gained by participating increases.
3. The actors who gain more utility will become powerful.
4. Powerful actors are interested in an important DCFM. Also, important DCFMs grab the attention of powerful actors.
5. An actor needs control over the event to become powerful. Also, a powerful actor would assert greater control over the outcome of the DCFM.

We developed these relations more rigorously as postulates and mathematical formulations below. Deviant cyber flash mob (DCFM) is an instance of cyber collective action (CCA), and could have either of the two outcomes, i.e., success (1) or failure (0) to achieve its goal [44]. To determine whether a DCFM will succeed or not, we propose five postulates that capture the relations between the factors identified above.

Postulate 1: If the Importance (Im) of DCFM increases, then the interest (I) of the actors to participate increases. Using the logical implication symbol ($\rightarrow$), this relation can be expressed as:

$$Importance(Im) \rightarrow Interest(I)$$

If the number of interested (I) actors in a DCFM increases, then the importance (Im) of the DCFM increases. Using the logical implication symbol and logical equivalence symbol ($\leftrightarrow$), this relation can be expressed as:

$$Interest(I) \rightarrow Importance(Im)$$

$$\therefore Importance(Im) \leftrightarrow Interest(I) \tag{1.1}$$

Postulate 2 If the amount of utility (U) gained by participating increases, then the interest (I) of an actor in the DCFM will increase. This relation can be expressed as:

$$Utility(U) \rightarrow Interest(I)$$

Postulate 3 If the amount of utility (U) gained by participating in the DCFM increases, then the actor's power (P) will increase. This relation can be expressed as:

$$Utility(U) \rightarrow Power(P)$$

Postulate 4 If the actor is powerful (P), then (s)he will be interested in important (Im) DCFMs. This can be expressed as:

$$Power(P) \rightarrow Importance(Im)$$

If the DCFM is important (Im), then powerful (P) actors will be interested in participating. This relation can be expressed as:

$$Importance(Im) \rightarrow Power(P)$$

$$\therefore Power(P) \leftrightarrow Importance(Im) \tag{1.2}$$

Postulate 5 If the actor is powerful (P), then (s)he will assert more control (C) on the DCFM. This relation can be expressed as:

$$Power(P) \rightarrow Control(C)$$

If the actor asserts more control (C) on the DCFM, then (s)he will be powerful (P). This can be expressed as:

$$Control(C) \rightarrow Power(P)$$

$$\therefore Power(P) \leftrightarrow Control(C) \tag{1.3}$$

The above postulates lead to the following formulations:

$$Power(P) = f(C, Im)$$

$$Importance(Im) = f(Interest)$$

$$Interest(I) = |U_{outcome1} - U_{outcome0}|$$

$$\therefore Power(P) = f(C, |U_{outcome1} - U_{outcome0}|)$$

Or,

$$\therefore Power(P) = f(C, I) \tag{1.4}$$

We posit that in order for a DCFM to succeed it needs to have many powerful actors interested in outcome (1). The sum of the powers of all the actors that are interested in the DCFM will give the amount of importance of that DCFM. If that amount of the summation of power equals to a threshold value (which can be calculated from the historical data of known DCFMs that succeeded in the past), then it will be possible to determine whether a DCFM is going to succeed or not.

$$Im = \sum_{1}^{m} P_j \geqslant Threshold\ Value \tag{1.5}$$

The summation of powers will be a scalar value that shows how many actors (nodes) control the DCFM and are interested in the outcome 1 (because their utility will increase). Next, we discuss the two cases of success and failure of DCFM in more depth.

1.3.1 The Case of DCFM-Success

To model the formation of a DCFM, it is essential to understand the source of motivation of the individuals that coordinate the act. Shared orientations among individuals often form the basis for motivation resulting in collective actions [71, 72]. Shared orientations among individuals induce a sense of belongingness to the group giving rise to the group's collective identity. Social dimensions [73], such as affiliations, interests, time, location, among others, are the shared orientations that govern the relationships among individuals coordinating a DCFM. Individuals may be connected along one or more social dimension, resulting in multiple shared orientations and hence a stronger collective identity. It is important to note that such relations are supra-dyadic[13] and multidimensional that are best modeled using hypergraph[14] [75]. Simple graphs (explained in Sect. 2.1), although efficiently capture the dyadic relations among a set of nodes, are severely challenged in modeling supra-dyadic and multidimensional relations commonplace in DCFM settings [74].

To illustrate these limitations of simple graphs, let's consider a collaboration network represented using a simple graph, where nodes denote authors and edges denote collaborations between the authors. Such a representation would tell us whether any two authors have collaborated or not. However, we cannot know whether three or more authors connected to each other collaborated on the same article. A bipartite graph (explained in Sect. 2.1) can possibly be used to address this limitation by creating two different sets of nodes: one denoting authors and the other denoting articles. The edges connecting the nodes across the two sets would denote the collaboration relationship among authors. However,

1. Such a representation does not allow studying network properties that require homogeneity among nodes, e.g., connectivity, centrality, and other structural/topological properties.
2. Such a representation only allows us to model the situations where relations among the actors are governed by a single common process. Relationships governed by multiple processes such as the relationships between actors coordinating an act based on time and location of the event and their affiliations could not be modeled by a single bipartite graph. Using n-partite graphs to model n-processes governing relations between actors is a possible but extremely complicated and unscalable solution.

So for such a complicated system we represented it using hypergraph notation because simple and bipartite or n-partite graph would not be sufficient to represent multidimensional and supra-dyadic relations that this system has [76, 77].

---

[13] Supra-dyadic relations are the relations that involve more than two nodes, i.e., a set of nodes such as food webs [74].

[14] Hypergraph is a graph in which each edge/relation is called a hyperedge and it connects more than two nodes that can be of different types.

1.3.2   The Case of DCFM-Failure

In his book entitled, "The Logic of Collective Action," Mancur Olson [49] put forward a single basic premise of collective action: "…individual rationality is not sufficient for collective rationality…" [78]. Olson's classic book [49] is mostly concerned with explaining and illustrating how *collective failure* results when individuals pursue self-interest. Olson's argument is essentially based on the assumption that every individual acts rationally, but if the individuals as a group choose not to act rationally—with respect to individual costs and benefits—*no collective action would occur*. So, if an actor **has control** on DCFM, but **does not have an interest** in the success of DCFM, then the actor has two choices, i.e., (s)he either not participates from the DCFM or asserts power exchange with other members of DCFM (to gain control over other events (DCFMs) or to gain social capital). By doing so the social capital of the individual will increase, while the power on that DCFM might decrease. Social capital is defined as "the value that one gains from personal connections such as membership in a family, an ethnic association, elite clubs, or other solidarity groups," as stated by the French economic and cultural sociologist Pierre Bourdieu [44, 47, 79, 80]. It should be noted that an actor may lose interest in the DCFM's success, if (s)he does not perceive any gain in utility by acting in the DCFM or face a risk.

If some of the actors in the DCFM are **not interested** in participating and they **do not have control** over the outcome, then they will have two choices, i.e., either they will not participate or will not participate and act against the group [44]. DCFMs can have two possible models: *a flat model* when the benefits/utilities of the action are equally distributed between all the actors (nodes) who are participating in the collective action. Or, DCFMs could assume *a hierarchical model* for distribution of the benefits/utilities of the action. In such a hierarchical model, benefits/utilities are disproportionately distributed among the group members. Further, the higher the rank of the member, the more benefits/utilities the member has. Imaginably, withdrawal of actors from a DCFM with a hierarchical model (especially at the top of the pyramid) will have a bigger effect than the withdrawal of the individual from a DCFM with a flat model.

1.3.3   Conceptual Framework

First, DCFM data need to be collected from online social networks. Second, this data need to be explored to find out what kind of information we can extract about the users (i.e., identify the shared orientations between group members like social dimensions such as their interest, location, time, and affiliations). Third, this data need to be explored to find out what kind of information we can extract about the DCFM itself (i.e., the number of users involved in that flash mob, whether it succeeds or not, etc.). The more the shared orientations that exist among the members of the group, the stronger is the collective identity [81] because shared orientations among individuals induce a sense of belongingness to the group giving rise to the group's collective identity. Melucci [82] argues that the collective identity formation is the intermediate process for the manifestation of contemporary forms of collective actions in the information age.

After that we will use hypergraph to represent this complex system. Formally, a hypergraph is a generalization of a graph, where an edge can connect any number of vertices. A hypergraph H is a pair $H = (X, E)$, where X is a set of nodes or vertices, and E is a set of non-empty subsets of X called hyperedges [83]. A simple graph can be considered as a special case of hypergraph, where each hyperedge has a cardinality of 2. So let A be the incidence matrix $(m \times n)$ of the social network (DCFM data that contain users and the relations between them), where rows represent nodes (vertices) and columns represent hyperedges (relations). Matrix A will be used to identify the motivation factors (shared orientations "Social Dimensions"). For each node we need to determine whether it is interested in outcome 1 or outcome 0 (the question whether the node's utility will increase or decrease by each outcome). The utility of the actors (nodes) of the deviant cyber flash mob can be estimated from:

1. Historical data (Examine the nodes of the deviant cyber flash mob historically participated in previous DCFM. If they did participate in that flash mob that indicates a strong *interest* of those nodes in acting in the current cyber flash mob and that their *utility* will increase by participating.),
2. Clustering coefficient (The higher the clustering coefficient, the more that node shares common orientation and is willing to increase its utility), and
3. Utility is also social media platform dependent (i.e., which platform the actor is using), e.g., an actor can gain more utility if (s)he gain a lot of retweets, or likes on a Facebook status or images, etc.

Let Y be the *relative utility difference matrix* $(m \times n)$ of the network where y_{ji} will have positive value if actor j favors participation based on hyperedge i. Similarly y_{ji} will have negative value if actor j favors not to participate based on hyperedge i. U_{j1i} is the utility of actor j gained by participating in the DCFM from hyperedge i. U_{j0i} is the utility of actor j gained by not participating in the DCFM from hyperedge i, such that:

$$y_{ji} = \frac{U_{j1i} - U_{j0i}}{\sum_i |U_{j1i} - U_{j0i}|} \tag{1.6}$$

Let X represent the interest matrix $(m \times n)$ of the individuals in the network. Since the interest of an actor in a flash mob is his utility difference from both the outcomes; therefore, x_{ji} will be

$$x_{ji} = y_{ji} \tag{1.7}$$

From the interest matrix X now we know how many individuals are interested in outcome 1 (e.g., positively interested) and how many individuals are interested in outcome 0 (e.g., negatively interested).

We also need to calculate the control (C_{ji}) of each node on the CFM. We use the control matrix C ($m \times n$) to represent that and C_{ji} values can be obtained by using the eigenvector centrality (centrality measures are explained in Sect. 2.2.1) [75]

$$Centrality\ measure\ of\ individuals\ C_N = AA^T \qquad (1.8)$$

$$Centrality\ measure\ of\ hyperedges\ C_E = A^T A \qquad (1.9)$$

Now the importance of a CFM [84] which is a function of the control and interest can be obtained using the following algebraic expression:

$$Importance\ (Im) = \sum_1^m P_j \geqslant Threshold\ Value$$

In matrix notation, when we have a system of m-equations:

$$Importance\ (Im) = X.C_N \qquad (1.10)$$

1.3.4   A CFM Scenario

To study the proposed model's efficacy, we present a CFM scenario and with the help of a toy example demonstrate how to assess the outcome of the CFM. Let the CFM scenario be—on March 15, 2018 at 11:00 am, all students who study at *University Z*, their age is more than 30, they live in *City, State Y*, and have red cars go to *Location L* parking lot which is located in *City, State X*, and dance in the parking lot. Each participant will get $300 for participating in this event.

We assume that this event (CFM) has only two outcomes (binary):

1. Participate, i.e., Success, or 1.
2. Do not participate, i.e., Fail, or 0.

Also, this CFM has two hyperedges (i):

1. Cost
2. Time

Actor j will have utility if he participates, i.e., his utility for outcome 1 or for participation. Actor j will also have utility by not participating, i.e., his utility for outcome 0 or for not participating.

Case 1:   The CFM will cost actor j money and actor j will not make it on time

- Actor j's utility if he decided to participate (1) in this CFM:

  1. Assuming the trip will cost actor j \$500, and he is getting \$300 for participating, then (\$300 − \$500 = \$−200). So actor j will lose \$200 since it's a negative cost.
  2. Assuming actor j will wake up at 9:00 am, and it will take him 5 h to go to the CFM, then (9:00 am + 5 h = 2:00 pm, 11:00 am to 2:00 pm = - 3 h). So actor j will miss the CFM by 3 h.

- Actor j's utility if he decided not to participate (0):

  1. Actor j will not lose \$200, so his cost is \$0.
  2. Actor j will not go, so he will not reach the event after 3 h, so the time he lost is 0 h.

The relative utility difference will be *negative* because:

$$y_{jcost} = \frac{U_{j1cost} - U_{j0cost}}{\sum_{cost} |U_{j1cost} - U_{j0cost}|}$$

$$y_{jcost} = \frac{-200 - 0}{\sum_{cost} |-200 - 0|}$$

$$y_{jcost} = -1$$

$$y_{jtime} = \frac{U_{j1time} - U_{j0time}}{\sum_{time} |U_{j1time} - U_{j0time}|}$$

$$y_{jtime} = \frac{-3 - 0}{\sum_{time} |-3 - 0|}$$

$$y_{jtime} = -1$$

So,

$$Y = \begin{bmatrix} & j & .. & .. & V_n \\ Cost & -1 & .. & .. & .. \\ Time & -1 & .. & .. & .. \\ & .. & .. & .. & .. \\ & .. & .. & .. & .. \end{bmatrix}$$

Case 2:   Actor j will make money by participating in the CFM and he will make it on time

- Actor j utility if he decided to participate (1) in this CFM:

  1. Assuming the trip will cost actor j \$200, and he is getting \$300 for participating, then (\$300 – \$200 = \$100) so he will gain \$100 since it's a positive cost.
  2. Assuming actor j will wake up at 5:00 am, and it will take him 5 h to go to the CFM, then (9:00 am + 5 h = 10:00 am, 11:00 am to 10:00 pm = + 1 h, i.e., he will be there 1 h before the event time and will not miss the CFM).

- Actor j utility if he decided not to participate (0):

  1. He will not gain \$100, but his cost is still \$0.
  2. He will be bored and miss all the fun, but the time he lost is 0 h.

In this case the relative utility difference will be *positive* because:

$$y_{jcost} = \frac{U_{j1cost} - U_{j0cost}}{\sum_{cost} |U_{j1cost} - U_{j0cost}|}$$

$$y_{jcost} = \frac{100 - 0}{\sum_{cost} |100 - 0|}$$

$$y_{jcost} = +1$$

$$y_{jtime} = \frac{U_{j1time} - U_{j0time}}{\sum_{time} |U_{j1time} - U_{j0time}|}$$

$$y_{jtime} = \frac{1 - 0}{\sum_{time} |1 - 0|}$$

$$y_{jtime} = +1$$

So,

$$Y = \begin{bmatrix} & j & \cdots & V_n \\ Cost & +1 & \cdots & \cdots \\ Time & +1 & \cdots & \cdots \\ & \cdots & \cdots & \cdots \\ & \cdots & \cdots & \cdots \end{bmatrix}$$

From the *interest matrix* or the *relative utility difference matrix*, we know that actor j in *Case 1* has a positive interest or will gain utility and in *Case 2* he has a negative interest or will lose utility. We also need to calculate the *control* actor j has on the outcome of the CFM. In both of the cases presented above, if actor j has a lot of *control* on the outcome of the CFM, let's assume he is one of the organizers of the CFM, then he has more control value as an actor who is only a hobbyist participant. In this toy example, the control of actor j can be estimated by the number of friends that their participation is depending on actor j participation in the CFM (e.g., the more dependent friend's actor j has the more control actor j has on the success of the CFM). Using the four scenarios presented in Sect. 1.3.2 we can determine how many actors are going to participate in the CFM. By doing so for many CFM of the same category, a predictive model based on historical data of various CFMs can be built.

The model presented above and demonstrated through a toy example leveraged theories and concepts that have been studied in social science to study or understand deviant behaviors conducted in cyberspace, i.e., social media. With the advent of this rapidly growing social space, there is a need to revisit various social science theories and assess their applicability to the cyberspace, then build tools that leverage these models and theories to help authorities develop strategies to counter such behaviors and enhance overall cyber operations. This model should help understand and advance cyber security strategies at a fundamental social and behavioral level. The research is of potential interest to sociologists, anthropologists, and information system experts exploring the influence of social systems on user behaviors; studying ties between people, technology, and institutions; and examining organizational structures, roles, and crowd processes.

In the next chapter, we provide fundamental concepts and terminologies about social network analysis (SNA). The readers need to know these in addition to the problem of deviance in social media (explained in this chapter) to understand the opportunities that social cyber forensics (SCF) (explained in Chap. 4) can bring to solve some of these problems/challenges.

References

1. *Data Never Sleeps 5.0 | Domo*. Available: https://www.domo.com/learn/data-never-sleeps-5?aid=ogsm072517_1&sf100871281=1
2. B. Marr, *How Much Data Do We Create Every Day? The Mind-Blowing Stats Everyone Should Read*. Available: https://tinyurl.com/yclva46f
3. N. Agarwal, M. Lim, and R.T. Wigand, *Online Collective Action: Dynamics of the Crowd in Social Media* (Springer, Vienna, 2014)
4. N. Agarwal et al., Raising and rising voices in social media: a novel methodological approach in studying cyber-collective movements. Bus. Inf. Technol. Syst. Eng. **4**(3), 113–126 (2012)
5. N. Staff, The Arab Spring: A year of revolution, in *National Public Radio (NPR)* (Dec. 2011). Available: http://n.pr/1EjLbeQ
6. M. Atkinson, Parkour, anarcho-environmentalism, and poiesis. J. Sport Soc. Issues **33**(2), 169–194 (2009)
7. Y.S. Mohilever, Taking over the city: developing a cybernetic geographical imagination-flash mobs & parkour, in *Theatre Space After 20th Century* (2012) p. 188
8. "EEAS - European External Action Service - European Commission. Available: https://eeas.europa.eu/headquarters/headquarters-homepage_en

9. EU vs Disinformation - EU vs Disinformation. Available: https://euvsdisinfo.eu/
10. Definition of Deviance in English by Oxford Dictionaries. Available: https://en. oxforddictionaries.com/definition/deviance
11. Deviance. Available: https://www.thefreedictionary.com/deviance
12. J. Devine, F. Egger-Sider, Beyond Google: the invisible web in the academic library. J. Acad. Librariansh. **30**(4), 265–269 (2004)
13. S. Raghavan, H. Garcia-Molina, Crawling the Hidden Web, Stanford, Tech. Rep. (2000)
14. N. Hamilton, The mechanics of a deep net metasearch engine, in *WWW (Posters)* (2003)
15. A. Greenberg, *Hacker Lexicon: What is The Dark Web?* (Nov. 2014). Available: https://www. wired.com/2014/11/hacker-lexicon-whats-dark-web/
16. M.K. Bergman, White paper: the deep web: surfacing hidden value. J. Electron. Publ. **7**(1), (2001)
17. S. Al-khateeb, M.N. Hussain, N. Agarwal, Social cyber forensics approach to study twitter's and blogs' influence on propaganda campaigns, in *International Conference on Social Computing, Behavioral-Cultural Modeling and Prediction and Behavior Representation in Modeling and Simulation* (Springer, Berlin, 2017), pp. 108–113.
18. S. Al-khateeb, K.J. Conlan, N. Agarwal, I.Baggili, F. Breitinger, Exploring Deviant Hacker Networks (DHN) on social media platforms. J. Digit. Forensic Secur. Law **11**(2), 7–20. Available: http://bit.ly/2nKwNJE
19. D. Sindelar, *The kremlin's troll army: Moscow is financing legions of pro-Russia Internet commenters. But how much do they matter? The Atlantic (Aug. 2014).* Available: http://www. theatlantic.com/international/archive/2014/08/the-kremlins-troll-army/375932/
20. K. Vick, *Iraq's Second Largest City Falls to Extremists.* Available: http://ti.me/1ElCjnV
21. T. Arango, *Key Iraqi City Falls to Isis as Last of Security Forces Flee.* Available: http://nyti. ms/1Tm4SOc
22. T. A. et al., *How ISIS Works* (2014). Available: http://nyti.ms/1h0spmR
23. M. Townsend, T. Helm, *Jihad in a Social Media Age: How Can the West Win an Online War?.* Available: http://bit.ly/1wqeb4b
24. J. Berger, How ISIS Succeeds on Social Media Where #StopKony Fails Even with fewer clicks. The Atlantic (March 16, 2015)
25. S. Al-khateeb, N. Agarwal, Analyzing deviant cyber Flash Mobs of ISIL on twitter, in *Social Computing, Behavioral-Cultural Modeling, and Prediction* (Springer, Berlin, 2015), pp. 251– 257.
26. C. Staff. *ISIS Video Appears to Show Beheadings of Egyptian Coptic Christians in Libya.* Available: http://www.cnn.com/2015/02/15/middleeast/isis-video-beheadings-christians/
27. T.n. editorial. *ISIL Executes an Israeli Arab After Accusing him of Been an Israeli Spy.* Available: http://www.tv7israelnews.com/isil-executes-an-israeli-arab-after-accusing-him-of-been-an-israeli-spy/
28. K. Shaheen. *ISIS Video Purports to Show Massacre of Two Groups of Ethiopian Christians.* Available: https://www.theguardian.com/world/2015/apr/19/isis-video-purports-to-show-massacre-of-two-groups-of-ethiopian-christians
29. C. News, ISIS recruits fighters through powerful online campaign, in http://cbsn.ws/ 1qKH8mE. Last checked: July 1, 2015, August 29, 2014
30. D. Quiggle, The ISIS beheading narrative. Small Wars J. (Feb 26, 2015). Accessed 12 Mar 2019. Available: https://smallwarsjournal.com/jrnl/art/the-isis-beheading-narrative
31. R. Janes, *Losing Our Heads: Beheadings in Literature and Culture* (NYU, New York, 2005)
32. T.J. Holt, Examining the forces shaping cybercrime markets online. Soc. Sci. Comput. Rev. **31**(2), 165–177 (2013). Available: http://ssc.sagepub.com/content/31/2/165.short
33. T.J. Holt, Exploring the social organisation and structure of stolen data markets. Glob. Crime **14**(2–3), 155–174 (2013). Available: http://www.tandfonline.com/doi/abs/10.1080/17440572. 2013.787925
34. E. Moreau, *Here's What you Need to Know About Internet Trolling.* Available: https://www. lifewire.com/what-is-internet-trolling-3485891

35. Z. Davis, *Definition of: Trolling* (2009). Available: http://www.pcmag.com/encyclopedia/term/53181/trolling#
36. I. University, *What is a Troll?* (2013). Available: https://kb.iu.edu/d/afhc
37. M. Allen, *Kremlin's 'Social Media Takeover': Cold War Tactics Fuel Ukraine Crisis* (Mar. 2014). Available: http://www.demdigest.net/blog/kremlins-social-media-takeover-cold-war-tactics-fuel-ukraine-crisis/
38. Oxford-Dictionary, Definition of flash mob from Oxford English dictionaries online, in *Oxford English Dictionaries*. https://en.oxforddictionaries.com/definition/us/flash_mob. Last checked 19 Dec 2016
39. C. Kirkland, *12 Great Examples of Flash Mobs* (Dec. 2011). Available: http://econsultancy.com/blog/8548-12-great-examples-of-flash-mobs?utm_campaign=bloglikes&utm_medium=socialnetwork&utm_source=facebook
40. G. Ackerman, D. Blair, G. Butler, H. Cabayan, R. Damron, J.D. Keefe, T. King, D. Hallstrom, S. Helfstein, D. Hulsey, et al., *The New Face of Transnational Crime Organizations (TCOs): A Geopolitical Perspective and Implications to US National Security* (Calhoun Institutional Archive of the Naval Postgraduate School, Mar. 2013). https://calhoun.nps.edu/public/handle/10945/30346.
41. B. Holbrook, *LBPD Prepared For Potential Bash Mob Event* (Jul. 2013), p. 00000. Available: http://www.everythinglongbeach.com/lbpd-prepared-for-potential-bash-mob-event/
42. S. Al-khateeb, N. Agarwal, Analyzing flash mobs in cybernetic space and the imminent security threats a collective action based theoretical perspective on emerging sociotechnical behaviors, in *2015 AAAI Spring Symposium Series*.
43. *Cyveillance, Bashmobs: Using Social Media to Organize Disruptive Activity*. Available: https://tinyurl.com/y9rlzrah
44. J.S. Coleman, The Mathematics of Collective Action. (New York, Routledge, July 12, 2017). 1st edn., 2005. 246p, eISBN 9781351479714. Accessed 12 Mar 2019. Available: https://www.taylorfrancis.com/books/9781351479714; https://doi.org/10.4324/9781315133065
45. B. Klandermans, J.M. Sabucedo, M. Rodriguez, M. De Weerd, Identity processes in collective action participation: Farmers' Identity and farmers' protest in the Netherlands and Spain. Polit. Psychol. **23**(2) 235–251. Available: http://onlinelibrary.wiley.com/doi/10.1111/0162-895X.00280/abstract
46. A. Melucci, *Challenging Codes: Collective Action in the Information Age* (Cambridge University, Cambridge, 1996), p. 00017
47. J.S. Coleman, *Foundations for a Theory of Collective Decisions*, vol. 71(6). Available: https://tinyurl.com/y7cugwp7
48. V. Labatut, N. Dugue, A. Perez, *Identifying the Community Roles of Social Capitalists in the Twitter Network*, p. 8. Available: http://arxiv.org/abs/1406.6611
49. M. Olson, *The Logic of Collective Action: Public Goods and the Theory of Groups* (Harvard University, Cambridge, 1977)
50. R.H. Coase, The nature of the firm. Economica **4**(16), 386–405 (1937)
51. P.G. Warr, Pareto optimal redistribution and private charity. J. Public Econ. **19**(1), 131–138 (1982)
52. P.G. Warr, The private provision of a public good is independent of the distribution of income. Econ. Lett. **13**(2), 207–211 (1983)
53. T. Sandler, J.T. Tschirhart, The economic theory of clubs: an evolutionary survey. J. Econ. Lit. Am. Econ. Assoc. **18**(4), 1481–1521 (1980)
54. A. Rubinstein, Equilibrium in supergames with the overtaking criterion. J. Econ. Theory **21**(1), 1–9 (1979)
55. G.S. Becker, *The Economic Approach to Human Behavior* (University of Chicago, Chicago, 1976)
56. M.N. Zald, J.D. McCarthy, *The Dynamics of Social Movements: Resource Mobilization, Social Control, and Tactics* (Winthrop, Cambridge, 1979)
57. H.R. Kerbo, Movements of crisis and movements of affluence a critique of deprivation and resource mobilization theories. J. Confl. Resolut. **26**(4) 645–663 (1982)

58. M.M. Ferree, in Frontiers in Social Movement Theory, ed. by A.D. Morris, C.M. Mueller (Yale University, Yale, 1992)(2), pp. 29–52.
59. D.A. Snow, et al., Frame alignment processes, micromobilization, and movement participation. Am. Sociol. Rev. **51**(4) 464–481 (1986)
60. H. Johnston, E. Larana, J.R. Gusfield, Identities, grievances, and new social movements, in *New Social Movements: From Ideology to Identity*, vol. 3, (Temple University Press, Philadelphia, PA, 1994), pp. 3–35. Available: https://www.jstor.org/stable/j.ctt14bst9g
61. S. Al-khateeb, N. Agarwal, Developing a conceptual framework for modeling deviant cyber flash mob: A socio-computational approach leveraging hypergraph constructs. J. Digit. Forensic. Secur. Law **9**(2), 113–128 (2014)
62. A. Cheng, M. Evans, *Inside Twitter an In-Depth Look at the 5% of Most Active Users.* Available: http://sysomos.com/insidetwitter/mostactiveusers
63. "*Types of Bots: An Overview of Chatbot Diversity |botnerds.com.* Available: http://botnerds. com/types-of-bots/
64. R.A. Rodríguez-Gómez, G. Maciá-Fernández, P. García-Teodoro, Survey and taxonomy of botnet research through life-cycle. ACM Comput. Surv. **45**(4), 45. Available: http://dl.acm.org/ citation.cfm?id=2501659
65. A. Karasaridis, B. Rexroad, D. Hoeflin, Wide-scale botnet detection and characterization, in *Proceedings of the First Conference on First Workshop on Hot Topics in Understanding Botnets*, Cambridge MA, vol. 7. . Available: http://bit.ly/2n4fmXk
66. S. Bono, D. Caselden, G. Landau, C. Miller, Reducing the attack surface in massively multiplayer online role-playing games. IEEE Secur. Priv. **7**(3) (2009)
67. N. Abokhodair, D. Yoo, D.W. McDonald, Dissecting a social botnet: Growth, content and influence in twitter, in *Proceedings of the 18th ACM Conference on Computer Supported Cooperative Work & Social Computing* (ACM, New York, 2015), pp. 839–851. Available: http://dl.acm.org/citation.cfm?id=2675208
68. P. Kulp, *Facebook Admits to Nearly as Many Fake or Clone Accounts as the U.S. Population 2017.* Available: https://goo.gl/HXUdc8
69. M. Newberg, Nearly 48 Million Twitter Accounts Could be Bots, Says Study (Mar. 2017). Available: https://www.cnbc.com/2017/03/10/nearly-48-million-twitter-accounts-could-be-bots-says-study.html
70. Y. Boshmaf, I. Muslukhov, K. Beznosov, M. Ripeanu, Key challenges in defending against malicious socialbots, in *Proceedings of the 5th USENIX Conference on Large-Scale Exploits and Emergent Threats* (USENIX Association, Berkeley, 2012), pp. 12–12. Available: https:// www.usenix.org/system/files/conference/leet12/leet12-final10.pdf
71. A. Melucci, *Nomads of the Present: Social Movements and Individual Needs in Contemporary Society* (Temple University, Philadelphia, 1989) p. 02740
72. L. Maheu, *Social Movements and Social Classes: The Future of Collective Action.* SAGE Studies in International Sociology, vol. 46 (SAGE Publications Ltd, London, 1995), p. 00620. Available: http://bit.ly/2nKwsqC
73. L. Tang, H. Liu, Toward predicting collective behavior via social dimension extraction. IEEE Intell. Syst. **25**(4), 19–25 (2010). Available: http://bit.ly/2n4mT8w
74. T. Shafie, D. Schoch, J. Mans, C. Hofman, U. Brandes, Hypergraph representations: a study of Carib attacks on colonial forces, 1509–1700. J. Hist. Netw. Res. **1**(1), 52–70 (2017)
75. P. Bonacich, A. Cody Holdren, M. Johnston, Hyper-edges and multidimensional centrality. Soc. Netw. **26**(3), 189–203 (2004). Available: http://linkinghub.elsevier.com/retrieve/pii/ S0378873304000024
76. E. Estrada, J.A. Rodriguez-Velazquez, Subgraph centrality and clustering in complex hyper-networks. Physica A Stat. Mech. Appl. **364**, 581–594 (2006)
77. P. Bonacich, et al., Hyper-edges and multidimensional centrality. Soc. Netw. **26**(3), 189–203 (2004)
78. T. Sandler, *Collective Action: Theory and Applications* (University of Michigan, Ann Arbor, 1992)
79. N.W. Biggart, *Readings in Economic Sociology*, vol. 4 (Blackwell, Chichester, 2002)

80. N.B. Ellison, C. Steinfield, C. Lampe, The benefits of Facebook "friends:" social capital and college students' use of online social network sites. J. Comput. Mediat. Commun. **12**(4), 1143–1168. Available: http://doi.wiley.com/10.1111/j.1083-6101.2007.00367.x
81. B. Klandermans et al., Identity processes in collective action participation: farmers' identity and farmers' protest in the Netherlands and Spain. Polit. Psychol. **23**(2), 235–251 (2002)
82. A. Melucci, *Challenging Codes: Collective Action in the Information Age* (Cambridge University, Cambridge, 1996)
83. C. Berge, E. Minieka, *Graphs and Hypergraphs*, vol. 7 (North-Holland, Amsterdam, 1973)
84. P. Ludlow, What is a 'Hacktivist'?. Available: http://opinionator.blogs.nytimes.com/2013/01/13/what-is-a-hacktivist/

Chapter 2
Social Network Measures and Analysis

Abstract In this chapter, we present basic terminologies and concepts of *graph theory* in addition to a few social network measures that will be used throughout the book. Then we explain more advanced metrics and concepts that would leverage the basic network measures such as estimating *blogs* and *bloggers'* influence scores and focal structures analysis (FSA). These concepts were used in many real-world cases to find coordinating sets of individuals (coordinating groups) in a given graph. All the concepts and measures are described and illustrated with examples. This chapter would provide the readers with basic understanding of graph-theoretic concepts and social network measures that will help understand the concepts of social cyber forensics in the later chapters.

Keywords Graph theory · Graph data structures · Social network analysis (SNA) · Centrality measures · Clustering coefficient · Modularity · Influence

2.1 Basics of Graph Theory

Graph theory is a mathematical theory that studies graphs and models pairwise relations between objects. The concept of a graph was first introduced in the eighteenth century by the Swiss mathematician Leonhard Euler as he solved the problem of $K\ddot{o}nigsberg$[1] bridge using a graph. In that problem, he presented the seven bridges of $K\ddot{o}nigsberg$ by edges and the land separated by these bridges as vertices (see Fig. 2.1). Then he proved that it is impossible to cross all the bridges once and only once in a single trip by starting and ending in the same point [1, 2]. Since then, graph theory is being used to model and study many processes and relations in social, physical, biological, and information systems. In sociology, graphs can be used to measure an individual's prestige, spread of rumors or happiness [3, 4], or community detection [5], in biology, graphs can be used to study

---

[1] $K\ddot{o}nigsberg$ is a city formerly a part of Germany and now it is a part of Russia known as Kaliningrad.

© The Author(s), under exclusive license to Springer Nature Switzerland AG 2019   27
S. Al-khateeb, N. Agarwal, *Deviance in Social Media and Social Cyber Forensics*,
SpringerBriefs in Cybersecurity, https://doi.org/10.1007/978-3-030-13690-1_2

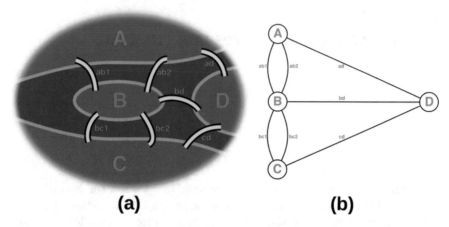

Fig. 2.1 (a) *Königsberg* bridge problem map, (b) representation of the *Königsberg* bridge problem as a graph

the breeding patterns or the spread of diseases [6, 7], in computer science, graphs can represent the flow of information between different computational devices, and in chemistry, graphs can be used to represent a molecule also known as "molecular graphs" [8].

A graph consists of points, or nodes, or vertices that are connected by relations, links, or edges. Mathematically, a graph G is defined as pair $G(V, E)$, where V represents the set of vertices and E represents the set of edges [9]. The graph vertices can represent various objects such as entities including actors, physical locations, organizations, political affiliation, etc., while the lines or arcs represent relations or linkages between these vertices. The words "Graph,[2]" "Network,[3]" and "Sociogram[4]" all refer to the same concept but they are called differently depending on the context and the discipline in which they are used.

Graphs have various types depending on *the types of vertices*, and *types of edges*. Graphs can be directed or undirected, i.e., whether each edge has an orientation/direction or not. Undirected graphs can be considered as a special case of a directed graph, in which every edge can be split into two directed edges in either directions. Directed graphs are also called "Digraph." Often "Graph" simply means undirected graph. A graph can also have weight w on each edge, hence these are called "Weighted graphs." Weighted graphs can be used to model things like distance, delay, bandwidth, the importance of an edge, etc. Unweighted graph is a special case of weighted graph with $w(u, v) = 1$ for all edges. Figure 2.2 shows three graphs, the one on the left is an example of an unweighted and undirected

---

[2]A graph consists of vertices connected by edges.

[3]A network consists of nodes connected by links.

[4]A sociogram consists of actors/points connected by relations.

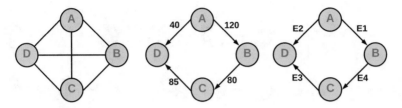

Fig. 2.2 Graph on left is an unweighted and undirected graph. Graph in the middle is a weighted and directed graph. Graph on the right is an unweighted and directed graph

graph that consists of vertices connected by line segments (edges), such that a graph $G = (V, E)$ with vertices represented by $V = \{A, B, C, D\}$ and edges represented by $E = \{(A, B), (B, C), (C, D), (D, A)\}$, while the one on the right represents the opposite case, i.e., a weighted and directed graph. Graph in the middle is a weighted and directed graph. *Note that a graph can also be unweighted and directed or weighted and undirected.*

A simple graph is a graph in which all vertices have the same type and are connected at most by one edge type, e.g., a Twitter user connected to another Twitter user by an edge of type "friend." A multigraph is a graph in which all vertices have the same type and are connected by multiple edges, e.g., a Twitter user connected to another Twitter user by two edges of type "friend" and "retweet." In multigraphs vertices are not allowed to connect to itself (i.e., self-loops are not allowed), but if a vertex connected to itself, then it is called a "pseudograph". For example, in Twitter network a user can be connected to another user by two type of edges "friend" and "retweet," in addition to that the same user if tweeted, then that tweet can be represented as a self-loop edge of type "tweet." Figure 2.3 shows the three types of graphs mentioned above. Note that the edges or connections between vertices have different colors which can represent different types of edges, e.g., in Fig. 2.3 black can represent *friends*, blue can represent *retweets*, red can represent *mentions*, and orange can represent *tweets*.

A graph can also have different types of vertices, e.g., a vertex of type "person" is connected to a vertex of type "company," and both are connected to a vertex of type "location." These types of graphs are called K-partite graphs.[5] In K-partite graphs, vertices can be divided into "K disjoint sets so that no two vertices within the same set are adjacent" [10]. If graph contains two types of vertices it is called "bipartite graph" (or 2-partite, see the left graph in Fig. 2.4) and if it has three types of vertices it is called "tripartite graphs" (or 3-partite graph, see the middle graph in Fig. 2.4). Sometimes K-partite graphs are also called "multimode graphs." If a graph consists of various types of vertices that are connected by various types of edges it

---

[5]K refers to the number of vertices types in the graph.

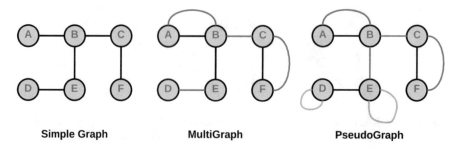

Fig. 2.3 Graph on the left represents a simple graph, graph in the middle represents a multigraph, and graph on the right represents a pseudograph

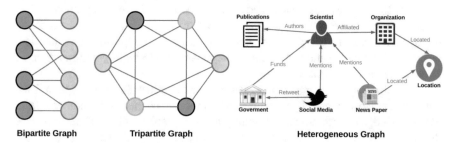

Fig. 2.4 Graph on the left is a bipartite, graph in the middle is a tripartite, and graph on the right represents a heterogeneous graph

is called a "heterogeneous graph," e.g., in academic environment publications are connected to scientists, scientists are connected to organizations, and a scientist can also be mentioned on social media [11]. The figure on the right in Fig. 2.4 shows a heterogeneous graph of various entities connected via various relations.

In addition to *the types of vertices*, and *types of edges* a graph can have various names depending on the *number of edges* it has. A graph is called a "complete graph" or a "clique" if each vertex is connected to every other vertex. A complete and undirected graph with N vertices will always have $((N^2 - N)/2)$ edges.[6] Figure 2.2 on left is an example of a complete and undirected graph. A graph can also have many "subgraphs." A subgraph of a graph $G = (V, E)$ is a graph $G' = (V', E')$ whose vertex set V' is a subset of V and edge set E' is a subset of E restricted to vertices in V'. A supergraph of a graph G' is a graph in which G' is a subgraph. These concepts are extremely helpful in defining communities of users in social networks.

---

[6]Q. How many edges are there in a complete and directed graph with N vertices? Ans: $(N^2 - N)$.

2.1.1 Graph Data Structures

In this section, we introduce the different forms of graph data structures along with examples.

2.1.1.1 List Structure

Graphs can be represented using the list data structure which will present a graph as a "List." This list can be a list of edges which is called an **incidence list**. Figure 2.2 on left can be represented as an incidence list in this format:

$$\{(AB/BA), (BC/CB), (CD/DC), (AD/DA), (AC/CA), (BD/DB)\}$$

and Fig. 2.2 on right can be represented as an incidence list in this format:

$$\{(AB, 120), (AD, 40), (BC, 80), (CD, 85)\}$$

Graphs can also be represented as a list in which each vertex in the graph has a list of directly connected vertices. This type of graph representation is called an **adjacency list**. Figure 2.2 on left can be represented as an adjacency list in this format:

$$\{(A : B, D, C), (B : A, C, D), (C : B, D, A), (D : C, B, A)\}$$

and Fig. 2.2 on right can be represented as an adjacency list in this format:

$$\{(A : B(120), D(40)), (B : C(80)), (C : D(85))\}$$

2.1.1.2 Matrix Structure

Graphs can be represented using a matrix data structure which will represent a graph as a "matrix." This matrix is called an **incidence matrix** if vertices are inserted as rows and edges are inserted as columns. Each entry of the incidence matrix represents $[vertex, edge]$ and will have a value of 0 or 1 to denote if the edge is incident upon the vertex or not [12]. Matrix on left in Fig. 2.5 is the incidence matrix representation of the right graph in Fig. 2.2.

Graphs can also be represented using another type of matrices which is called the **adjacency matrix** or the **sociomatrix**. An adjacency matrix is an $n \times n$ matrix, where n is the number of vertices in a graph. Each entry of the adjacency matrix represent $[vertex(i), vertex(j)]$ and will have a value of 0 or 1 to denote if there is an edge from vertex (i) to vertex (j) [12]. Matrix in the middle of Fig. 2.5 is the adjacency matrix representation of the right graph in Fig. 2.2.

	E1	E2	E3	E4
A	-1	-1	0	0
B	1	0	0	-1
C	0	0	-1	1
D	0	1	1	0

Incidence Matrix

	A	B	C	D
A	0	1	0	1
B	0	0	1	0
C	0	0	0	1
D	0	0	0	0

Adjacency Matrix

	A	B	C	D
A	0	1	2	1
B	∞	0	1	2
C	∞	∞	0	1
D	∞	∞	∞	0

Distance Matrix

Fig. 2.5 On left an incidence matrix, in the middle an adjacency matrix, and on right a distance matrix all for the graph on right in Fig. 2.2

Finally, a graph can also be represented as a **distance matrix**, which is an $n \times n$ matrix, where n is the number of vertices in a graph. Each entry of the distance matrix represents the shortest path (explained in Sect. 2.2) between $vertex(i)$ and $vertex(j)$ and will have various values to denote the number of edges on the shortest path between $vertex(i)$ and $vertex(j)$. Matrix on right in Fig. 2.5 is the distance matrix presentation of the right graph in Fig. 2.2. Note that when $vertex(i) = vertex(j)$ the shortest path is 0 and when $vertex(i)$ and $vertex(j)$ are not reachable the shortest path is ∞ or undefined.

Using a list data structure or a matrix data structure to represent graphs has both advantages and disadvantages, e.g., list data structure is memory efficient, used by most visualization and analysis tools, but it has slower lookup which can be improved by using hash index structures. On the other hand, matrix data structure is memory inefficient, challenging for large graphs (especially big social networks), but it has faster lookup.

2.2   Fundamentals of Social Network Analysis (SNA)

In this section, we provide a set of basic and advanced social network measures that are derived from graph theory. These set of measures are necessary for the reader to understand before advancing to other chapters.

A social network is a network of social entities that are connected by various social ties, for example, members of a group who are friends, live in the same zip code, and go to the same school form a social network. Social network analysis (SNA) focuses on studying the relationships among these social entities [12], identify various patterns, and understand the implications of these ties. SNA is vastly used in various disciplines such as industrial engineering, economics, social science, and behavioral science to study various topics such as the political and economic system of the world [13, 14], occupational mobility [15], and formation of coalitions [16–18].

A social network can be represented as a graph with vertices (social entities) and edges (social ties). If these vertices are connected, i.e., there is a sequence of *distinct vertices and edges*, then this sequence is called a "path" [12]. The length of the path (also known as the number of hops) is equal to the number of edges traversed between the origin and destination vertices.[7] The "geodesic" distance is the *shortest path* between any two vertices in a graph and it is defined as $d(v_i, v_j)$, where v_i and v_j are the two vertices in the graph. The path between vertex A and vertex D in the left graph of Fig. 2.2 can be $ABCD$ which has a length of 3 or the path can be AD which has a length of 1. The latter case is the geodesic between A and D.

If the sequence has *only distinct edges and some repeated vertices*, then it's called a "trail," e.g., $ACBAD$ is a trail between vertex A and vertex D for the left graph in Fig. 2.2. On the other hand, if the sequence has *non-distinct vertices and edges*, then it is called a "walk," e.g., $ABCABD$ is a walk between vertex A and vertex D for the left graph in Fig. 2.2. Note that:

- every path can be considered as a trail but not every trail can be considered as a path,
- every trail can be considered as a walk but not every walk can be considered as a trail, and
- every path is a walk but not all walks are paths [12].

One of the important graph-theoretic measures is "connectedness," which is a graph property that indicates whether a graph is connected or not. A graph is considered as connected if every vertex is reachable from every other vertex, i.e., there is a path between every pair of vertices. If there exists a node that is not reachable by any other node, then the graph is "disconnected." The maximal connected subgraph is called "component." If a graph contains one component, then it is *connected* and if it has more than one component it is called *disconnected*. Directed graph or a digraph is *strongly connected* if every vertex is reachable from every other vertex following the direction of the edges, and it is *weakly connected* if the graph obtained after ignoring edge directions is connected.

Graph diameter is a measure that represents the largest distance between any pair of vertices in the graph [12]. The diameter value can range from a minimum of 1 (if the graph is complete such as the graph in the right in Fig. 2.2) to a maximum of $n - 1$, e.g., graph (b) in Fig. 2.1. If the graph is not connected, then the diameter is undefined (or infinite). The diameter can be calculated for graph components/subgraphs and its value will reflect the maximum geodesic between vertices within the subgraph.

---

[7]The small world experiment conducted by the American social psychologist Stanley Milgram in 1967 leverage this concept. In his experiment, he measured the average path length between people in the USA. He found that on average any two randomly selected people living in the USA are connected by 5.5 (or, 6) hops [19]. The phrase "six degrees of separation" is associated with his experiment although he didn't use this phrase.

2.2.1   Centrality Measures

There are many graph-theoretic measures that can be applied on a graph (a set of vertices and edges) to analyze and derive insights. One of the most fundamental measure is the degree of a vertex, $deg(v)$, e.g., the user's number of friends on Facebook. The degree of a vertex or "**degree**" is defined as *the number of edges a vertex has*. For example, each node in the graph in the left of Fig. 2.2 has a degree of 3. If a vertex has no adjacent vertices, then its degree = 0 and it is called an isolate (short for an isolated vertex).

Directed graphs have "**indegree**" and "**outdegree**"[9]. Indegree represents the number of in-coming edges to a vertex $deg(v^{in})$, while outdegree represents the number of out-coming edges from a vertex $deg(v^{out})$ such as the number of a Twitter user's followers and followees, respectively. In the graph in the middle of Fig. 2.2 vertex A has an outdegree of 2 and an indegree of 0, while vertex D has an outdegree of 0 and an indegree of 2. In directed graphs, degree of a vertex is the sum of its indegree and outdegree, i.e., $deg(v) = deg(v^{in}) + deg(v^{out})$. To make degree, indegree, and outdegree comparable across graphs, its values need to be normalized (i.e., between 0 and 1). Normalized degree measures are called *degree centralities*. Degree centrality, indegree centrality, and outdegree centrality for normalized degree, indegree, and outdegree, respectively. These measures are interesting because they can be used to estimate "popularity" or "receptivity" (from indegree centrality) and "expansiveness" or "gregariousness" (from outdegree centrality). There are three main ways to normalize:

1. Normalization by maximum possible degree (divide by $n - 1$, where n is the total number of vertices in a graph), hence degree centrality in this case will be $C_d^{normal}(v_i) = \frac{deg(v_i)}{n-1}$.
2. Normalize by the maximum degree (divide by the highest degree values of all the vertices in the graph), hence degree centrality in this case will be $C_d^{normal}(v_i) = \frac{deg(v_i)}{deg^{max}(v_j)}$, where $deg^{max}(v_j)$ is the highest degree values of all the vertices in the graph.
3. Normalize by the degree sum (i.e., divide by the sum of all vertices' degrees OR divide by $2 \times m$, where m is the total number of edges in the graph), hence degree centrality in this case will be $C_d^{normal}(v_i) = \frac{deg(v_i)}{2 \times m}$.

Other social statuses in a social graph can be estimated using other centrality measures such as **betweenness centrality**, closeness centrality, and eigenvector centrality.

Betweenness centrality for vertices measures the number of shortest paths between other nodes that pass through that vertex. Vertices that occur on many shortest paths between other vertices have higher betweenness than those that do not. Such vertices are important in connecting other vertices in a graph and act as information "brokers" or "bridges". Mathematically, *betweenness centrality* is defined as:

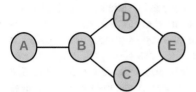

Nodes	Betweenness Centrality	Closeness Centrality
A	0.0	0.5
B	0.58	0.8
C	0.16	0.66
D	0.16	0.66
E	0.08	0.57

Fig. 2.6 On left an undirected graph and on right a table of each vertex betweenness and closeness centralities

$$C_B^{normal}(v) = \frac{\sum_{s \neq v \neq t \in V, s \neq t} \frac{\sigma_{st}(v)}{\sigma_{st}}}{(n-1)(n-2)} \tag{2.1}$$

where σ_{st} is the number of geodesic paths (shortest paths) between the source vertex (s) and the target vertex (t). $\sigma_{st}(v)$ is the number of those geodesic paths passing through vertex (v). To make the vertex betweenness centrality comparable across graphs, its values need to be normalized by the *maximum value* it takes. A vertex will have a maximum betweenness centrality if it lays on all geodesic paths between any source (s) and target (t) for any pair (s,t) that is $[(n-1)(n-2)]$, e.g., vertex B on the left side of Fig. 2.6 has the highest betweenness centrality, while vertex A has a betweenness centrality value of 0 since no geodesic paths pass through it. Below we demonstrate how the betweenness centrality measure is calculated for nodes C and D:

$$n = 5, so(n-1)(n-2) = 12$$

$$C_B^{normal}(C) = \frac{2 \times \left(\underbrace{0}_{s=A,t=B} + \underbrace{0}_{s=A,t=D} + \underbrace{(1/2)}_{s=A,t=E} + \underbrace{0}_{s=B,t=D} + \underbrace{(1/2)}_{s=B,t=E} + \underbrace{0}_{s=D,t=E} \right)}{12}$$

$$C_B^{normal}(C) = \frac{2 \times 1.0}{12} = 0.16$$

$$C_B^{normal}(C) = C_B^{normal}(D) = 0.16$$

Note that the values are multiplied by 2 because it is an *undirected graph*. The table on the right in Fig. 2.6 shows the betweenness centrality values of all other vertices in the left graph.

Betweenness centrality shows the importance of a vertex in connecting other vertices, in addition to that, how quickly a vertex can reach other vertices makes it central in a graph. This can be calculated using the **closeness centrality**, which is defined as the mean geodesic (g) distance between a vertex (v) and all other reachable vertices. Centrally located vertex has a small distance to all other vertices in the graph. In other words, the vertex with the smallest average shortest path has the highest closeness centrality. Vertices that tend to have short geodesic distances to other vertices within the graph have higher closeness, e.g., vertex B in Fig. 2.6 on left has the highest closeness centrality, while vertex A has the lowest closeness centrality. Mathematically, closeness centrality is defined as:

$$C_C(v_x) = \frac{1}{\frac{\sum_{v_y \neq v_x} st_{v_x,v_y}}{(n-1)}} \tag{2.2}$$

where (st_{v_x,v_y}), is the geodesic path from vertex v_y to vertex v_x. Below we demonstrate how the closeness centrality measure is calculated for node A:

$$C_C(A) = \frac{1}{\frac{1+2+2+3}{(5-1)}} = \frac{1}{\frac{8}{(4)}} = 0.5$$

The table on the right in Fig. 2.6 shows the closeness centrality values of all other vertices in the graph on the left. Each of the aforementioned centrality measures has their own way of calculating how central a vertex is for the graph; hence, a central vertex in one measure might not be as central according to the other centrality measure.

There are other variations of centrality measure, e.g., Eigenvector Centrality, Katz Centrality, and PageRank (an improvement of Katz Centrality), that are helpful in determining how central/important a vertex is in a graph. Interested readers can refer to [9] for more details about these centrality measures. The aforementioned centrality measures incorporate the importance of the vertex neighbors (in an ego-alter network[8]) to the vertex own importance or centrality. For example, connecting to more "popular" individuals is not the same as connecting to less "popular" ones. We cover a variety of social network analysis tools in Chap. 3 that provide easy to use interfaces to compute these centrality scores for a social network.

2.2.2 Triadic Closure and Clustering Coefficient

The concept of *Triadic closure* was first suggested by the German sociologist and philosopher Georg Simmel in 1908 [21]. In his book, "Sociology: Investigations on the Forms of Sociation," he suggested the idea of triadic closure between three vertices in a graph, i.e., if two people know a common person, then they are more

---

[8] An ego-alter network is a network in which you are called the "ego" and your friends are called the "alters" [20].

Fig. 2.7 Two different
triplets, same nodes
but missing different edges

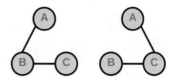

likely to know each other in future. This likelihood increases with more common
individuals. One way to measure triadic closure is by calculating the *clustering
coefficient* of a graph.

There are two types of clustering coefficients. The first type is called the **global
clustering coefficient** which was introduced by Luce and Perry in 1949 [22]. The
global clustering coefficient gives a measure for the whole graph (i.e., graph-level
analysis). The global clustering coefficient is based on *triplets* of vertices in a graph.
A triple is a set of three ordered vertices that are connected by edges. If the three
vertices are connected by three edges, then it is called a *closed triplet* and if they are
connected by two edges, then it is called an *open triplet*. The triplets will be different
if they have different nodes or have the same nodes but different missing edges, for
example, in Fig. 2.7 the triplet $V_A V_B V_C$ is not the same as the triplet $V_A V_C V_B$
because they are missing edges $E(V_A V_C)$ and $E(V_A V_B)$, respectively. The global
clustering coefficient is defined as:

$$GC_{CO} = \frac{\text{(Number of closed triplets)}}{\text{(Number of all connected triplets)}} \qquad (2.3)$$

$$GC_{CO} = \frac{\text{(Number of } (T_C))}{((\text{Number of } (T_C)) + (\text{Number of } (T_O)))} \qquad (2.4)$$

where (T_C) stands for closed triplets and (T_O) stands for open triplets. The number
of closed triplets (T_C) is equal to the number of closed path of length 2. Since every
triangle can have six closed paths of length 2, then the global clustering coefficient
can be defined as:

$$GC_{CO} = \frac{3 \times \text{(Number of triangles)}}{(3 \times \text{(Number of triangles)} + (\text{Number of } (T_O)))} \qquad (2.5)$$

The left graph in Fig. 2.8 has a global clustering coefficient of 0.75. Below, we
explain how this was calculated:

$$GC_{CO} = \frac{3 \times \text{(Number of triangles)}}{(3 \times \text{(Number of triangles)} + (\text{Number of } (T_O)))}$$

$$GC_{CO} = \frac{3 \times (2)}{\left((3 \times (2)) + \underbrace{2}_{BAD,BCD}\right)} = 0.75$$

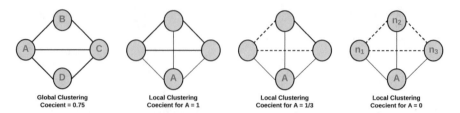

Fig. 2.8 Examples of global and local clustering coefficient. Solid lines depict connected vertices, while dashed lines depict missing connections among vertices

The other type of clustering coefficient is called the **local clustering coefficient** which was first introduced by Watts and Strogatz in 1998 [23]. The local clustering coefficient gives a measure for an individual vertex in the graph (vertex level analysis). This measure estimates how the neighbors of a vertex are close to being a clique (or a complete graph). The local clustering coefficient is defined as:

$$LC_{CO} = \frac{\text{(Number of pairs of neighbors for a vertex } V \text{ that are connected)}}{\text{(Number of all pairs of neighbors for a vertex } V)}$$

(2.6)

In undirected graphs, if vertex V has a K_i neighbors, then *the number of all pairs of neighbors for a vertex V* (the denominator) can be defined as: $\frac{K_i(K_i-1)}{2}$. The graph in the right of Fig. 2.8 has a local clustering coefficient of 0. Next, we explain how this was calculated. Since vertex A has three neighbors (n_1, n_2, *and* n_3) the denominator is

$$\frac{K_i(K_i - 1)}{2} = \frac{3(3 - 1)}{2} = 3$$

Since n_1 and n_2, n_1 and n_3, and n_2 and n_3 are not connected, the numerator = 0. Hence, the local clustering coefficient of vertex $A = 0$.

2.2.3   Modularity

Modularity is a network structural measure that evaluates the cohesiveness of a network [24, 25]. This measure is used to detect communities in a graph (a set of densely connected vertices within the group and sparsely connected with the rest of the graph). In other words, this measure defines how structured the community is, i.e., how far (how distant) their structure is from randomness. The assumption is real-world communities should be structured and not random [9].

Mathematically, for a graph $G(V, E)$, where each vertex degree is known in advance but not the edges themselves, and the number of edges in the graph $|E| = m$, considering two nodes, v_i with a degree of d_i and v_j with a degree of d_j, for any edges going out of v_i the probability this edge will go to v_j is $\frac{d_j}{2m}$. Since v_i has a degree of d_i, there are d_i number of edges. This means that the expected number of edges between v_i and v_j is $\frac{d_i d_j}{2m}$. If we partition the graph G into K partitions, then for each partition this distant or modularity is:

$$\sum_{v_i, v_j \in P_x} A_{ij} - \frac{d_i d_j}{2m}$$

But, since we have K partition in the graph, then the distant or modularity is:

$$\sum_{x=1}^{k} \sum_{v_i, v_j \in P_x} A_{ij} - \frac{d_i d_j}{2m}$$

For an undirected graph, all the edges are counted twice (i.e., $A_{ij} = A_{ji}$) and modularity [9, 25] is normalized and is defined as:

$$\text{Modularity } (Q) = \frac{1}{2m} \left(\sum_{x=1}^{k} \sum_{v_i, v_j \in P_x} A_{ij} - \frac{d_i d_j}{2m} \right) \qquad (2.7)$$

Figure 2.9 shows a small world[9] network of 50 vertices and 476 edges with a global clustering coefficient of 0.6 and local clustering coefficient (node average) of 0.592. The Newman modularity value—which gives a measure of the degree to which the grouping has found community structure—is 0.431. This network has three communities colored in red (8 nodes smallest community), blue (21 nodes), and green (21 nodes) in the left graph of Fig. 2.9, while these communities are collapsed to blocks view on the right. Identifying communities in a network is an important analysis to understand behaviors and derive insights, for example, identifying a community of closely connected individuals implies a faster rate of information (or rumor) dissemination than a loosely connected community.

2.2.4 Influential Blogs and Influential Bloggers

The popularity and ease of use nature of various social media platforms, e.g., Twitter, Facebook, VKontakte, YouTube, and blogs, provided a rich source of data for researchers in various disciplines, e.g., social, political, and behavioral

---

[9]Small world networks are known to have a high global clustering coefficient and an average shortest path length that increase slowly as the number of vertices increases. There are many examples of small world networks such as the electric power grids, networks of word co-occurrence, and the biological neural networks just to name a few.

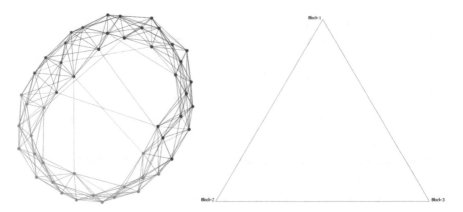

Fig. 2.9 A small world network with three communities on left. The same network presented on the right with block view of each community

sciences. Social media platforms have changed the way citizens express and share their opinions regarding various socio-political situations within their communities and have created a space for citizen journalism [26]. These new mediums of communication have also allowed for citizen opinions to be channelized from the online forum to the streets in the form of public debates and protests, e.g., during the Arab Spring [27] or the Venezuelan socio-political crisis and the migrant influx [28].

Blogs (short for weblog at a website [29]) are a rich-full source of textual data where narratives can be framed and disseminated. Blogs in combination with other social media platforms such as Twitter can be used to conduct deviant acts since it provides a rich medium for individuals to frame an agenda and develop discourse around it using half-truth or twisting facts to influence the masses. Twitter, however, due to the 280-character limit is primarily used as a dissemination medium. Bloggers use Twitter to build an audience (or, followership) and then drive their audience to their blogs in which they framed their narratives. It is important to understand the disinformation dissemination network on Twitter but it is equally, if not more, important to understand the blog environment and specifically the blogger's influence, engagement with the audience, and motivations for agenda setting.

Identifying influential individuals is a well-studied problem. Many studies have been conducted to identify the influence of a blogger in a community [29–33]. The basic idea of computing the influence a blogger has is to aggregate the influence of their individual blog posts. A blog post that has more *in-links* and *comments* indicates that the community is interested in it. In-links (indegree) and comments contribute positively towards the influence of the posts, whereas *out-links* (outdegree) of the blog posts contribute negatively towards the influence. Influence can be assessed using a stochastic model that uses *inbound links*, *comments*, and *outbound links* of the blog post as factors, as proposed in [29, 32].

Blog data is understudied and there is a lack of analytical tools[10] that does social network analysis on the blogosphere due to semi-structured nature of blog data and lack of application programming interface (API), etc. One of the tools that provide SNA on blogs data and can be used to estimate blogs' and bloggers' influence is Blogtrackers tool that will be introduced in Sect. 3.10.

2.2.5   Focal Structures Analysis (FSA)

Focal Structures Analysis [35] (FSA) is an advanced social network analysis methodology to discover an influential set of nodes or vertices in a network or graph. The nodes do not have to be strongly connected and may not be the most influential on their own but by acting together they form a compelling power. FSA algorithm has two versions:

- *Version 1:* The **Global Interconnected-based FSA**. It is a recursive modularity-based algorithm. This algorithm works in two steps: first it finds the candidate focal structures, then it will stitch those candidate structures together. The details of the steps are mentioned briefly below but interested readers are encouraged to read the full details of the algorithm in [35]. To find focal structures in a network, this algorithm follows the steps below:

  1. **Top-down division step:** The algorithm partitions the network into subgraphs or substructures. This obtains the candidate focal structures from the complex network by applying the Louvain method for computing modularity [36].
  2. **Bottom-up agglomeration step:** The FSA algorithm stitches the candidate focal structures, i.e., the highly interconnected focal structures or the focal structures that have the highest similarity values are stitched together, then the process iterates until the highest similarity of all sibling pairs is less than a given threshold value. The similarity between two structures is measured using the well-known Jaccard's coefficient [35, 37] which results in a value between 0 and 1 (where 1 means the two networks are identical and 0 means the two networks are not similar at all). The stitching of the candidate focal structures is performed to extract structures with low densities [35]. FSA can be freely used by anyone through the website (http://cosmos-1.host.ualr.edu/fsa)

---

[10]Most services are discontinued, e.g., Blogdex (developed by MIT and was shutdown in 2006), BlogPulse (developed by IntelliSeek and was shutdown in 2012), BlogScope (developed by the University of Toronto and was shutdown in 2012), Google Blog Search (BETA) (developed by Google and was shutdown in 2014), and Technorati (developed by Technorati and was shutdown in 2014) [34].

- *Version 2:* The **Local Interconnected-based FSA**. This version of FSA is a clustering coefficient-based algorithm [35, 38]. This version of FSA finds the focal structures in two steps:

  1. It starts by calculating the *local clustering coefficient* (explained in Sect. 2.2.2) for each vertex.
  2. Then it calculates the average of these values for all the vertices in the graph.
  3. Then it does pairwise comparisons for each pair of vertices. It compares their clustering coefficient values with the mean, if the value of both vertices clustering coefficient is greater than or less than the mean, then those vertices are put in the same focal structure [35, 38]. This version of FSA gives focal structures that contain vertices which are strongly connected and similar to each other.

The focal structures analysis has been evaluated on many real-world cases such as *the Saudi Arabian women's right to drive campaign "Oct26Driving campaign"* on Twitter [39]. In that study results showed that focal structures are *more interactive than average individuals* and *more interactive than communities* in the evolution of a mass protest. In other words, the interaction rate of focal structures is significantly higher than the average interaction rate of random sets of individuals. Also, the number of retweets, mentions, and replies increases proportionally with respect to the followers of the individuals within communities [35]. Interested readers are encouraged to read the full details of the study in [39].

The 2014 Ukraine Crisis is another example for FSA application. At that time, a British journalist and a blogger, Graham W. Phillips covered the 2014 Ukraine crisis and became a growing star on Kremlin-owned media [35, 40]. FSA was applied on a blog–blog network and the results illustrated that Graham Phillips was involved in the only focal structure of the entire network along with ITAR-TASS (the Russian News Agency) and Voice of Russia (the Russian government's international radio broadcasting service). Graham W. Phillips was actively involved in the crisis as a blogger and maintained a single-authored blog with significant influence when compared to some of the active mainstream media blogs [41].

The FSA approach was also applied to Twitter data during NATO's *Dragoon Ride 2015 Exercise* to discover the most influential set of bots (explained in Sect. 1.2.3) or the seeders of information in the graph. By applying FSA on the social network of these bots we obtained one focal structure, which contains two nodes. These two nodes form the most influential set of bots in the network, i.e., the bots that by working together profoundly impact the dissemination of propaganda [41].

In the next chapter, we introduce a set of tools and methodologies that can be used to collect and analyze data from various social media channels. These tools will help readers compute the aforementioned measures and be able to use social network analysis (SNA) to analyze data. Later, in Chap. 4 we will introduce the concept of social cyber forensics (SCF). We introduce a tool, i.e., Maltego, that can be used to perform SCF. This tool uses open source information (OSINF) to connect various entities which can help an analyst answer various questions. Both types of analysis, i.e., SNA and SCF, will be utilized in constructing a framework or methodology that we used to study various deviant cases introduced in Chap. 5. The set of analysis and tools introduced through the book will help readers have a better understanding of the problem of deviance in social media, and understand how both types of analysis can help in mitigating the effects of the problem on both cyber and physical spaces.

References

1. L. Mastin, *18th Century Mathematics - Euler* (2010), [Online]. Available: http://www.storyofmathematics.com/18th_euler.html
2. E.W. Weisstein, *Knigsberg Bridge Problem* (2018), [Online]. Available: http://mathworld.wolfram.com/KoenigsbergBridgeProblem.html
3. B. Doerr, M. Fouz, T. Friedrich, Why rumors spread so quickly in social networks. Commun. ACM **55**(6), 70–75 (2012)
4. J.H. Fowler, N.A. Christakis, Dynamic spread of happiness in a large social network: longitudinal analysis over 20 years in the Framingham heart study. BMJ **337**, a2338 (2008)
5. S. Fortunato, Community detection in graphs. Phys. Rep. **486**(3–5), 75–174 (2010)
6. O. Mason, M. Verwoerd, Graph theory and networks in biology. IET Syst. Biol. **1**(2), 89–119 (2007)
7. M.J. Keeling, K.T. Eames, Networks and epidemic models. J. R. Soc. Interface **2**(4), 295–307 (2005)
8. A.T. Balaban, Applications of graph theory in chemistry. J. Chem. Inf. Comput. Sci. **25**(3), 334–343 (1985)
9. R. Zafarani, M.A. Abbasi, H. Liu, *Social Media Mining: An Introduction* (Cambridge University Press, Cambridge, 2014)
10. E.W. Weisstein, *K-Partite Graph* (2018). [Online]. Available: http://mathworld.wolfram.com/k-PartiteGraph.html
11. M. Timilsina, I. Hulpus, C. Hayes, B. Davis, *Heterogeneous Graphs for Academic Impact Assessment*, Amsterdam, Netherlands (2015). [Online]. Available: http://elsevier.kdu.insight-centre.org/wp-content/uploads/altmetrics_2015_poster.pdf
12. S. Wasserman, K. Faust, *Social Network Analysis: Methods and Applications*, vol. 8 (Cambridge University Press, Cambridge, 1994)
13. D. Snyder, E.L. Kick, Structural position in the world system and economic growth, 1955–1970: a multiple-network analysis of transnational interactions. Am. J. Sociol. **84**(5), 1096–1126 (1979)
14. R.J. Nemeth, D.A. Smith, International trade and world-system structure: a multiple network analysis. Review (Fernand Braudel Center) **8**(4), 517–560 (1985)
15. L.A. Goodman, Criteria for determining whether certain categories in a cross-classification table should be combined, with special reference to occupational categories in an occupational mobility table. Am. J. Sociol. **87**(3), 612–650 (1981)
16. B. Kapferer, *Norms and the Manipulation of Relationships in a Work Context* (Manchester University Press, Manchester, 1969)
17. W.W. Zachary, An information flow model for conflict and fission in small groups. J. Anthropol. Res. **33**(4), 452–473 (1977)
18. B. Thurman, In the office: networks and coalitions. Soc. Networks **2**(1), 47–63 (1979)
19. J. Travers, S. Milgram, The small world problem. Phys. Today **1**(1), 61–67 (1967)
20. R. DeJordy, D. Halgin, *Introduction to Ego Network Analysis* (Boston College and the Winston Center for Leadership & Ethics, Boston, 2008)
21. G. Simmel, *Sociology: Investigations on the Forms of Sociation* (Duncker & Humblot, Berlin, 1908)

22. R.D. Luce, A.D. Perry, A method of matrix analysis of group structure. Psychometrika **14**(2), 95–116 (1949)
23. D.J. Watts, S.H. Strogatz, Collective dynamics of small-world networks. Nature **393**(6684), 440 (1998)
24. M. Girvan, M.E.J. Newman, Community structure in social and biological networks. Proc. Natl. Acad. Sci. **99**(12), 7821–7826 (2002). [Online]. Available: http://www.pnas.org/content/99/12/7821.abstract
25. M.E. Newman, Modularity and community structure in networks. Proc. Natl. Acad. Sci. **103**(23), 8577–8582 (2006). [Online]. Available: http://www.pnas.org/content/103/23/8577.short
26. D. Gillmor, We the media: the rise of citizen journalists. Natl. Civ. Rev. **93**(3), 58–63 (2004)
27. B. Etling, J. Kelly, R. Faris, J. Palfrey, Mapping the Arabic blogosphere: politics, culture, and dissent, in *Media Evolution on the Eve of the Arab Spring* (Springer, New York, 2014), pp. 49–74
28. E.L. Mead, M.N. Hussain, M. Nooman, S. Al-khateeb, N. Agarwal, Assessing situation awareness through blogosphere: a case study on Venezuelan socio-political crisis and the migrant influx, in *The Seventh International Conference on Social Media Technologies, Communication, and Informatics (SOTICS 2017)* (The International Academy, Research and Industry Association (IARIA), 2017), pp. 22–29
29. N. Agarwal, H. Liu, L. Tang, P.S. Yu, Identifying the influential bloggers in a community, in *Proceedings of the 2008 International Conference on Web Search and Data Mining* (ACM, New York, 2008), pp. 207–218
30. N. Agarwal, H. Liu, L. Tang, S.Y. Philip, Modeling blogger influence in a community. Soc. Netw. Anal. Min. **2**(2), 139–162. [Online]. Available: http://bit.ly/2mOw8HM
31. S. Kumar, R. Zafarani, M.A. Abbasi, G. Barbier, H. Liu, Convergence of influential bloggers for topic discovery in the blogosphere, in *Advances in Social Computing* (Springer, Berlin, 2010), pp. 406–412. [Online]. Available: http://bit.ly/2nxdHaA
32. A. Java, P. Kolari, T. Finin, T. Oates, Modeling the spread of influence on the blogosphere, in *Proceedings of the 15th International World Wide Web Conference* (2006), pp. 22–26. [Online]. Available: http://bit.ly/2nf4ZAA
33. K.E. Gill, How can we measure the influence of the blogosphere, in *WWW 2004 Workshop on the Weblogging Ecosystem: Aggregation, Analysis and Dynamics* (Citeseer, New York, 2004). [Online]. Available: http://bit.ly/2nt8bcs
34. M.N. Hussain, A. Obadimu, K.K. Bandeli, M. Nooman, S. Al-khateeb, N. Agarwal, A framework for blog data collection: challenges and opportunities, in *The IARIA International Symposium on Designing, Validating, and Using Datasets (DATASETS 2017)* (The International Academy, Research and Industry Association (IARIA), 2017)
35. F. Şen, R. Wigand, N. Agarwal, S. Tokdemir, R. Kasprzyk, Focal structures analysis: Identifying influential sets of individuals in a social network. Soc. Netw. Anal. Min. **6**(1), 1–22 (2016). [Online]. Available: http://bit.ly/1qS8Y4D
36. V.D. Blondel, J.-L. Guillaume, R. Lambiotte, E. Lefebvre, Fast unfolding of communities in large networks. J. Stat. Mech Theory Exp. **2008**(10), P10008 (2008)
37. P. Jaccard, The distribution of the flora in the alpine zone. New Phytol. **11**(2), 37–50 (1912)
38. F. Sen, N. Nagisetty, T. Viangteeravat, N. Agarwal, An online platform for focal structures analysis-analyzing smaller and more pertinent groups using a web tool, in *2015 AAAI Spring Symposium Series* (2015)
39. S. Yuce, N. Agarwal, R.T. Wigand, M. Lim, R.S. Robinson, Studying the evolution of online collective action: Saudi Arabian womens oct26drivingtwitter campaign, in *International Conference on Social Computing, Behavioral-Cultural Modeling, and Prediction* (Springer, Cham, 2014), pp. 413–420
40. M. Seddon, *How a British Blogger Became an Unlikely Star of the Ukraine Conflict and Russia Today*, (BuzzFeed News, New York, 2014). [Online]. Available: http://bzfd.it/1qpuL2z
41. S. Al-khateeb, N. Agarwal, R. Galeano, R. Goolsby, Examining the use of botnets and their evolution in propaganda dissemination. Def. Strateg. Commun. **2**(1), 87–112 (2017)

Chapter 3
Tools and Methodologies for Data Collection, Analysis, and Visualization

Abstract In this chapter, we briefly introduce a set of tools (mostly free and publicly available) and methodologies that can be used to collect, analyze, and visualize data from OSINF, Facebook, YouTube, Twitter, Blogs, and other sources. We will point out where you can get these tools, the capabilities of each tool, and how we used it in our research. This chapter is meant to give an overview of the tools currently used to conduct social network analysis (SNA), social cyber forensics (SCF), and text analytics. We will not cover the details of all the features/capabilities for each tool in this chapter—as some tools may require a whole book by themselves, but we will highlight each tool's importance and point out the available resources to interested readers. This chapter should help readers finding answers to some research questions using various tools for data collection, analysis, and visualization.

Keywords Data collection tools · Data analysis tools · Data visualization tools

3.1 TouchGraph SEO Browser

TouchGraph SEO[1] Browser is an online tool that can be used to search for words (e.g., a keyword or a hashtag) and visualize its mentions on the Internet. It returns the relevant websites as reported by Google database in a network view. This tool is available online at www.touchgraph.com/seo.

TouchGraph SEO Browser uses Java that can run on Mac OS, Windows, and Linux platforms. The tool visualizes returned Google search results for the keyword provided by the user. It will create an edge between the searched keyword and the domains that mention/contain the searched keyword. It also returns the number of times this keyword is "matched" (according to Google search algorithm) in that site.

---

[1]SEO stands for search engine optimization: which is the process of making sure that a website gets traffic by appearing in the search results of any search engine, e.g., Google, Bing, and Yahoo. This can be achieved by having the right content in the website in which search engines algorithms can use to make a site appear in the returned results when a search query is submitted.

The size of the circle around the domains icon is relative to the number of times the searched keyword is matched in that domain. For example, Fig. 3.1 show the results we obtain by inserting three keywords, i.e., "NATO," "USA," and "Russia," to the search bar of the tool. It shows that the top results returned by Google for the keywords "USA" and "NATO" are Twitter and YouTube while for "USA" and "Russia" it is www.cia.gov. The data used to generate the graph can be exported as .CSV file which can be further analyzed using other network visualization tools, introduced later in this chapter.

This tool is helpful when you want to collect data and have a list of keywords or hashtags. You can check how these keywords or hashtags are used in different social media channels or if it is even used at all in the social media channel you are interested in. The tool should help answer questions like: are the keywords of interest used more on Facebook or on Twitter?, which keyword is used on Facebook and which one is used on Twitter? answers to these types of questions will help in picking the right keywords for the right social media channel for a precise data collection [1].

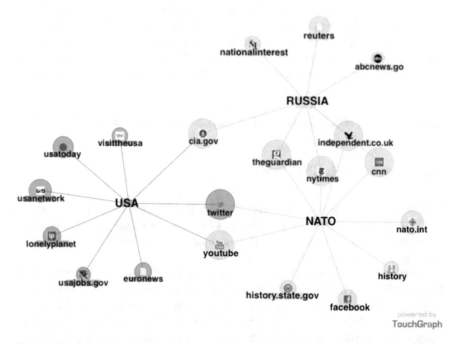

Fig. 3.1 TouchGraph results show how the three keywords "NATO," "USA," and "Russia" are connected through the respective domains as reported by Google search algorithm

3.2   Twitter Archiving Google Sheet (TAGS) and TAGSExplorer

Twitter Archiving Google Sheet (TAGS) is a Twitter archiving Google sheet developed by Martin Hawksey.[2] This tool can be used to collect Twitter search data and saves it to Google spreadsheet. The data collected can be up to 7 days old, i.e., TAGS can retrieve data from Twitter for the past 7 days prior to the date of the request. TAGS can also be set up to collect data from Twitter periodically, e.g., the sheet can be updated with new tweets every hour until you stop it. TAGS uses the Twitter REST APIs, i.e., GET search/tweets,[3] GET favorites/list,[4] and GET statuses/user_timeline[5] to collect Twitter data. The maximum number of returned tweets varies depending on the type of archive you are collecting. Data can be exported into .CSV format for further analysis with other tools.

TAGSExplorer is a part of TAGS. It can be used to visualize the collected data in a web browser (note that the spreadsheet has to be publicly viewable in order for TAGSExplorer to visualize it). It also gives some network measures such as top Tweeters (i.e., tweet a lot), top hashtags, and top conversationalist (i.e., mentioned in tweets and in replies a lot). Figure 3.2 shows the results we obtained by tracking

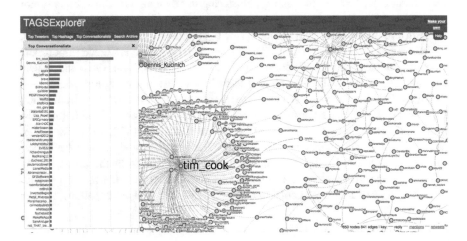

Fig. 3.2 TAGSExplorer results show @tim_cook as the top conversationalist in the collected data [2]

---

[2]Martin Hawksey is Chief Innovation, Technology and Community Officer at Association for Learning Technology, UK.

[3]https://dev.twitter.com/rest/reference/get/search/tweets.

[4]https://dev.twitter.com/rest/reference/get/favorites/list.

[5]https://dev.twitter.com/rest/reference/get/statuses/user_timeline.

the diffusion of #NoBackDoor on Twitter during Apple vs. FBI case.[6] Apple CEO @tim_cook was identified as top conversationalist because he was mentioned in many tweets and replies during the period February 16–May 21, 2016 [2].

3.3   Network Overview, Discovery, and Exploration for Excel (NodeXL)

NodeXL is an add-in for Microsoft Excel [3] developed by the Social Media Research Foundation[7] (SMR Foundation). It runs on Windows machines only and can be used with Microsoft Excel 2007 and up, to perform social network analysis (SNA) and content analysis [4]. NodeXL is available in two versions, *Basic* and *Pro*. The *Basic* version can be used to browse NodeXL files and it is available freely, while the *Pro* version is a paid version and provides features that enables users to collect and analyze different social media data.

NodeXL can be used to visualize networks with various layout algorithms and customizable nodes' or edges' properties such as colors, size, and labels. NodeXL can be used to perform social network analysis (SNA) such as calculating indegree, outdegree, density, modularity, clustering coefficient, pageRank, eigenvector centrality, closeness centrality, and betweenness centrality. It can also be used to perform content analysis such as identifying top words, bigrams (word pairs), URLs, hashtags, and time series analysis. The current version of NodeXL can import data from Facebook, YouTube, Twitter, Flickr, Pajek, UCINet, GraphML, and matrix formats. This tool is available online at http://www.smrfoundation.org/nodexl.

Figure 3.3, created using NodeXL, represents the social network or friends and followers of ISIL's top 10 propaganda disseminators which were identified in 2014 by the International Center for the Study of Radicalization (ICSR) [5]. The figure shows that ISIL disseminators follow each other but they rarely follow back their followers [1]. NodeXL is helpful when you do not want to reinvent the wheel, i.e., by writing a program to collect, analyze, and visualize Twitter data. It is also a powerful tool since it provides various content and social network analysis capabilities.

3.4   Gephi

Gephi is a tool for network analysis and visualization. It is developed by the Gephi Consortium which is a group of engineers and researchers in computer science. The tool is open source, freely available, and runs on Windows, Mac OS X, and Linux operation systems.

---

[6]Apple Right in Defying the F.B.I? The NY Times, 2016. Available at: http://nyti.ms/1qGtynH.
[7]SMR Foundation http://www.smrfoundation.org/research/our-network/.

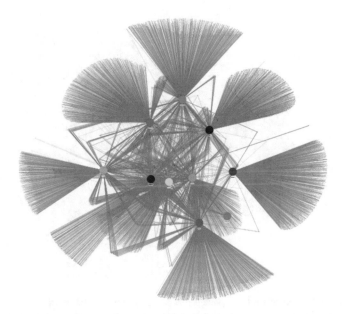

Fig. 3.3 The social network of ISIL's top 10 propaganda disseminators. Green edges are ISIL nodes following others. Red edges are other nodes following ISIL nodes. The nodes with different colors and bigger size represent ISIL top disseminators [1]

Gephi can be used to visualize networks with various layout algorithms and customizable nodes' or edges' properties such as colors, size, and labels. Gephi can be used to perform social network analysis (SNA) such as calculating the network diameter, shortest path, PageRank, modularity (for community detection), betweenness and closeness centralities, and clustering coefficient. Gephi provides dynamic filtering and can be used to analyze dynamic networks (temporal graphs) to observe how a network evolves over time. Gephi also has built-in "Plugins Center" which lists the available plugins from the Gephi plugin portal. This built-in plugins center extends the functionalities of Gephi, for example, the plugin "TwitterStreamingImporter" enables Gephi to collect data from Twitter Streaming API based on a keyword(s) or user screen name(s) then represent the collected data as a graph for further analysis. Three types of networks can be obtained using "TwitterStreamingImporter" plugin:

- Full Twitter Network: a graph consisting of users, hashtags, tweets, media, URLs, and their connection.
- Twitter User Network: a network of users and the relations between them.
- Twitter Hashtag Network: a network of co-occurring hashtags.

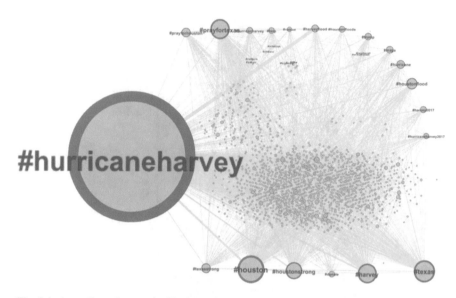

Fig. 3.4 An undirected network of hashtags that co-occur with the hashtags #hurricaneharvey and #prayfortexas. The nodes are represented in gray color, the edges in light green, and the node's labels in red color. The size of the node represents the degree centrality of the node [6, 7]

At the time of writing, the current version of Gephi (i.e., 0.9.1) can import data from CSV files, relational databases, and the majority of graph file formats such as DL files (UCINET) (*.dl), GraphViz files (*.dot and *.gv), Net files (Pajek), and other types. The network graphs created with Gephi can be exported to .PDF, .PNG, and .SVG. Gephi is available online at https://gephi.org.

We used the two hashtags, i.e., #hurricaneharvey and #prayfortexas to collect data using "TwitterStreamingImporter" plugin in Gephi during hurricane Harvey that hit the southeastern part of Texas in 2017. This plugin uses Twitter streaming API to get live data from Twitter. We made the plugin run for one hour, i.e., from 3:45 pm to 4:45 pm, US Central Time on Tuesday, August 29, 2017, and it collected 1891 nodes and 6128 edges. Figure 3.4 shows the network we obtained. In this figure, the nodes are the hashtags that co-occur with #HurricaneHarvey and #PrayForTexas. The edges represent co-occurrence relations. The hashtag network shows that the most co-occurred hashtags are asking people for help, staying positive and strong, prayers, the name of the hurricane, the name of the city and state effected by the hurricane, i.e., Houston, Texas, and #Trump as the US President Donald Trump visited Texas[8] on this date.

---

[8]Trump, in Texas, Calls Hurricane Harvey Recovery Response Effort a Real Team. The NY Times, 2017. Available at: https://www.nytimes.com/2017/08/29/us/trump-texas-harvey.html.

3.5 CytoScape

CytoScape is a tool that was originally designed for researchers in the Biology field to visualize and analyze molecular interaction, gene expression, and biological pathways. It is developed by CytoScape Consortium which is funded by the US National Institute of General Medical Sciences (NIGMS) and the National Resource for Network Biology (NRNB). The tool is open source, freely available, and runs on Windows, Mac OS X, and Linux operation systems. CytoScape can be downloaded from its website at http://bit.ly/1N2Rlrl.

CytoScape is nowadays used for complex network visualization and analysis with a basic set of features in addition to advanced features available through free "Apps" (formerly known as plugins). These apps equip CytoScape with new layouts, network and molecular profiling analyses, scripting, connection with databases, and additional file format support.

CytoScape is a good visualization tool as it provides easier control on the nodes' and edges' properties, e.g., color, width, shape, etc. We created Fig. 3.5 using CytoScape which represents a bipartite graph of Agent and Knowledge (i.e., Twitter User and Tweet) of bot accounts[9] that were disseminating ISIL's beheading propaganda videos on Twitter [8]. This figure shows that 18 bot accounts (the green ovals) are tweeting the same tweet[10] (the red triangle) at the same time.

3.6 Linguistic Inquiry and Word Count (LIWC)

Linguistic inquiry and word count (LIWC) is a very powerful tool for text analysis. It is developed by Pennebaker Conglomerates, Inc. LIWC can be tested for free using the following URL http://www.liwc.net/tryonline.php but to fully use it a license is required. LIWC has two types of license, i.e., "Academic" and "Commercial" (with API support). The academic license is for university or academic purposes only while the commercial license (provided by Receptiviti.ai) is for all other users. The tool works on Windows and Mac OS X operating systems.

LIWC accepts input format as .pdf, .rtf, and .csv and the output would be exported to a Microsoft Excel format, i.e., .csv, or tab-delimited file. The tool is available online at https://liwc.wpengine.com/.

---

[9]The bot accounts are: @DoneHealthE, @HealthNutre, @HealthDones, @Healthedone, @Health-Done1, @HealthDone3, @HealthFruit1, www.@Health_Years, @Fashion_DoneA, @Fashion_DoneB, @Fashion_DoneC, @SecretCode_1, @SecretCodeM, @SecretCodeM1, @Secret-CodeM2, @SecretCodeM4, @SecretCodeM5, @SecretCodeM6. Most of these accounts are suspended by Twitter.

[10]The tweet disseminated by the bot accounts is: ": Breaking: Isis beheaded 30 Ethiopian Christians in LibyaIsis beheaded 30 Ethiopian Christians in Libya [link to. . . ".

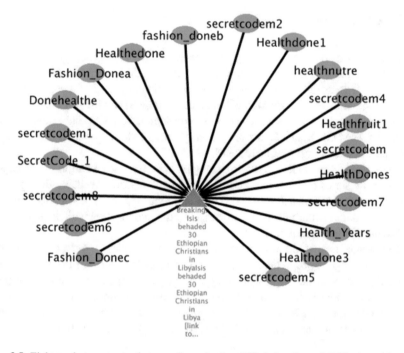

Fig. 3.5 Eighteen bot accounts that are disseminating ISIL beheading of Ethiopian video on Twitter. Green ovals are Twitter bot accounts, while the red triangle is the tweet. Edges in black color represent tweet relationship [8]

LIWC can do both linguistic and psychological analysis for a given text data. It gives a score for each of its dimensions. LIWC2015 has 16 main dimensions with a total of 82 sub-dimensions that can be translated into insights about the given corpus. These dimensions are: (1) *word count*, (2) *summary variables*, (3) *language metrics*, (4) *function words*, (5) *grammars*, (6) *affect words*, (7) *social words*, (8) *cognitive processes*, (9) *perpetual processes*, (10) *biological processes*, (11) *core drives and needs*, (12) *time orientation*, (13) *relativity*, (14) *personal concerns*, (15) *informal speech*, and (16) *all punctuation*.

We use LIWC to do text analysis of blog posts and save the results in a database that Blogtrackers tool (explained in Sect. 3.10) accesses and visualizes to derive insights [10]. Figure 3.6 shows the sentiment trends in Blogtrackers for 22 blog sites that were talking about the European migrant crisis[11] were thousands of refugees made their way to Europe from the Middle East and Africa in 2015.

---

[11]Migrant refugee crisis. BBC News December 17, 2015. Available at: http://bbc.in/2f7h3hr.

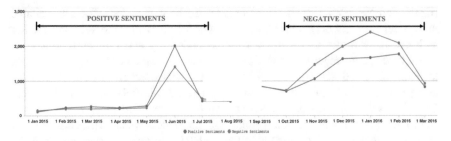

Fig. 3.6 Sentiment trend in the blogosphere during the European Migrant Crisis between January 2015 and March 2016 [9]

Fig. 3.7 (**a**) "Refugees Welcome" banners in a major soccer game, (**b**) "Rapefugees Not Welcome" banners in a street protest [9]

We collected 9183 blog posts from these blog sites during January 2015–March 2016. The sentiment trend was mostly positive from January 2015 to July 2015, neutral (positive and negative were equal) from July 2015 to October 2015, but after October 2015 there was a flip in the sentiments from positive to negative. The flip in sentiments, i.e., from a very positive sentiment towards the migrants—where citizens of many European countries sympathized and wanted their government to help the refugees, people raised "Refugees Welcome" banners at major soccer events (Fig. 3.7a)—to negative sentiments—where people protested in streets and raised "Rapefugees not Welcome" banners (Fig. 3.7b)—was mostly caused by the Paris attacks in November 2015 and the assault cases that were carried out by some immigrants in Germany during the new year's eve [9].

3.7   Organizational Risk Analyzer (ORA) NetScenes

Organizational risk analyzer (ORA) NetScenes is a toolkit for network analysis and visualization [11]. It is developed by the Center for Computational Analysis of Social and Organizational System (CASOS), at Carnegie Mellon University

(CMU) and Netanomics (a division of Carley Technologies, Inc). ORA is available in two versions ORA-Lite and ORA-Pro. ORA-Lite is the free version of ORA and can analyze and visualize up to 2000 nodes per entity types (entities can be of type: people/agent, knowledge, organization, resource, belief, event, task, location, role, and action). ORA-Lite can be downloaded from CASOS website using the following URL http://www.casos.cs.cmu.edu/projects/ora/versions.php. ORA-Pro is the paid version of ORA and has four types of license, i.e., individuals; academics; government and commercial; and distributors which can be downloaded from Netanomics website using the following URL http://netanomics.com/ora-pro/. ORA-Pro has no limits on the number of nodes and entity types that it can analyze and visualize. ORA-Pro works on both Windows and Mac OS X operating systems.

ORA has many network analysis measures that can be applied to different types of networks, e.g., multiplex,[12] multimode,[13] and dynamic networks (network that changes over time and space/location). ORA allows conducting social network analysis at different levels, i.e., *node, edge*, and *overall network*.

These measures are accessed through "Reports" that can be generated to give you an answer about a specific question, e.g., "Key Entity" report will identify the key actors based on their position in the network that is important to its operation. In addition to the network analysis capabilities that ORA has, it can also visualize data in 2D, 3D, and geo-data on different map layouts. ORA has an "Importer" which enables it to import data in a variety of formats such as Email data; data from a SQL database; Excel or text delimited files; XML network data such as DyNetML, GraphML, and Palantir XML; JSON data such as Twitter data, Twitter Stemmer data, Blogtrackers data, VKontakte (VK) data, another analysis tools, e.g., Pajek, UCINET (text or binary), PenLink, Analyst Notebook, Path Finder, Thing Finder, Personal Brain, and CmapTools, and other data formats such as Survey Monkey data, Shapefile data, Citations data, TAVI data, HIDTA data, and THINK data. Data can also be exported from ORA in a variety of formats such as DyNetML, Table format, Matrix, Node List, UCINET (text or binary), NetDraw (.vna), Pajek (.net), CmapTools, C3Trace, and Analyst Notebook.

ORA can be used for network analysis and visualization of a variety of datasets such as Twitter data, blogs data, and many others. Figure 3.8 shows the Twitter communication network of different *Blackhat hackers*. It's Twitter users network of mentions, retweets, and tweets. It contains 2740 nodes and 3445 edges. We started with a list of 49 deviant hackers known for their promotion of deviant activities. Note that some self-proclaimed "hacker" accounts on Twitter are actually benevolent in nature, i.e., "white-hat hackers" such accounts were avoided. We were able to identify 62 Twitter accounts for these deviant hackers (some groups have more than one Twitter account while others were not present on Twitter). Twitter API was used to collect their communication network, which is the network shown

---

[12]Multiplex is a network in which the nodes are connected via multiple types of relations/edges.

[13]Multimode is a network that has different types of nodes, e.g., 2-mode network, aka bipartite networks.

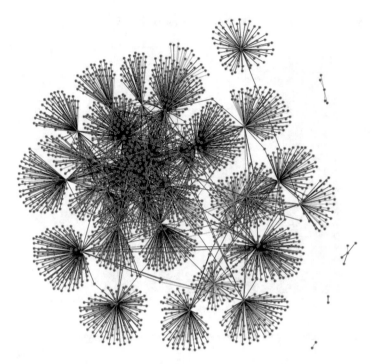

Fig. 3.8 The communication network of the deviant hackers on Twitter. Green edges represent "Retweets" relation. Black edges represent "Mentions" relation

in Fig. 3.8. This network was analyzed using ORA which helped us identify the accounts these deviant hackers interacted with the most, the accounts that tweeted the most, and the accounts that helped to spread the hacker's messages by retweeting it. Observing and monitoring this communication network of DHNs was key to understanding the content these Blackhat hackers were spreading, e.g., coordinating to conduct a cyber attack, operation information such as the name of the operation, attack strategy, and the outcome of the attack [12].

3.8   IBM Watson Analytics

IBM Watson Analytics is a cloud platform that can be used to analyze and visualize data in an interactive, intuitive, and quick manner. IBM Watson Analytics is available at https://www.ibm.com/us-en/marketplace/watson-analytics. It has three versions, *free*, *plus*, and *professional*. The difference between the three versions is the storage space and the number of users that can use the account. It's a cloud service so it is platform independent.

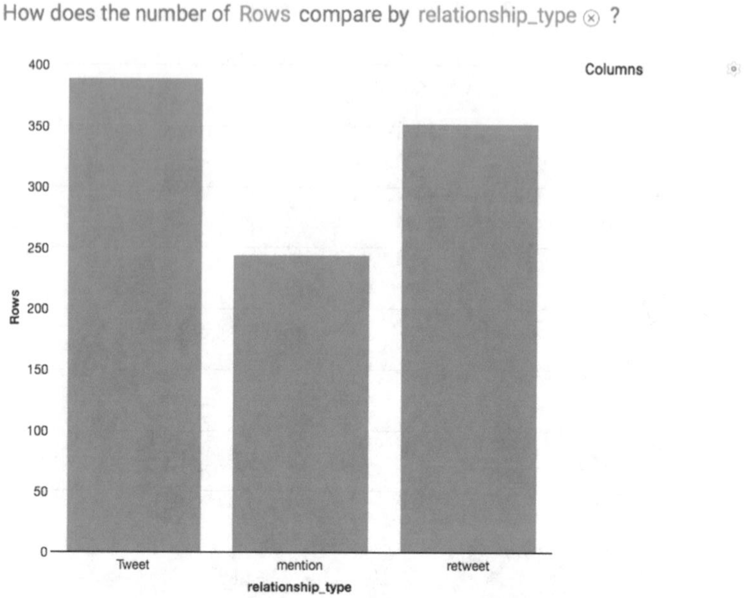

How does the number of Rows compare by relationship_type ⊗ ?

Fig. 3.9 The question along with its answer provided by Watson Analytics

IBM Watson Analytics is a powerful tool that makes smart data analysis and visualization very intuitive. It can help in discovering patterns in the data to derive insights. It has an "automated predictive analytics and cognitive capabilities" [13]. Data can be uploaded using spreadsheets, analyzed, and the findings can be stored in a visual dashboard. One very nice feature this tool has is the capability to interact with your data in a conversational manner, i.e., you can ask the tool questions which will be translated into a query then you get the results that would answer your question. For example, when we wanted to know how many tweets, mentions, and retweets are in one of the datasets we have collected for the case of Apple vs. FBI [2]. We asked Watson Analytics "How does the number of *rows* compared by *relationship type*?" it immediately gave us the results in a nice bar chart showing that it has 388 tweets, 351 retweets, and 243 mentions. Figure 3.9 shows the question along with its answer.

IBM Watson Analytics has five main modules: Explore, Predict, Assemble, Social Media, and Refine. These modules can do the following:

- The "Explore" module help the user find patterns and relationships among attributes in a given dataset by providing a conversational style questions template to help the user find the right question and obtain the right answer.

- The "Predict" module help the user understand what drives behaviors or outcomes in the data, i.e., by selecting an attribute you are interested in predicting its outcome/value it will use all the other attributes to estimate what are the contributing factors or attributes that will help achieve the outcome desired.
- The "Assemble" module would provide the user with the capability to organize all the findings in a nice visual dashboard.
- The "Social Media" module will help collect data from different social media channels such as Facebook and Twitter about a specific topic or keywords then allow the user to do all the other analysis on these datasets.
- The "Refine" module help the user shape and enrich the data to be used using Watson Analytics.

3.9 Web Content Extractor (WCE)

Web content extractor (WCE) is a software that can be used to extract data from websites and store it to a database of your choice. It is developed by Newprosoft. WCE has only one paid version[14] and it runs only on Windows machines.

Using WCE a user can train the crawler, i.e., identify the extraction pattern then WCE will do the rest of the job in an automatic manner. It is fast to learn and do not require any programming skills. It can run up to 20 simultaneous threads and uses many proxy servers to switch your machine IP address to avoid being blocked. Once the data is extracted, it can be exported to many data formats such as CSV, SQL, XML, TXT, HTML, Microsoft Access database, and to any Open Database Connectivity[15] (ODBC) source.

We used WCE extensively to crawl various blog sites then store the data to a MySQL database that Blogtrackers tool (explained in Sect. 3.10) can access and analyze. We have crawled more than 194 blog sites which were written in 41 languages and hosted on servers in 15 countries. These blogs were focused on different topics such as the Ukraine-Russia Conflict, anti-NATO narratives, the migrant crisis in the European Union, and the fake news blogs in the Baltic States [14]. Figure 3.10 shows the new project page in WCE. In this screen the user should insert the website of interest (the one to be crawled) then follow the steps to train the crawler on the extraction pattern, i.e., which element on the website needs to be crawled, e.g., the author, date, title of the page or article, etc. then where to store that information, i.e., which attribute, e.g., the author name will be stored in the attributes firstName and lastName. Once this is done, WCE will automatically crawl the website and all the pages linked (outbound links/URLs) in it.

---

[14]WCE can be downloaded from http://www.newprosoft.com/web-content-extractor.htm.

[15]ODBC is a protocol that is required to connect MS Access DB to MS SQL Server, https://goo.gl/9Wu2xy.

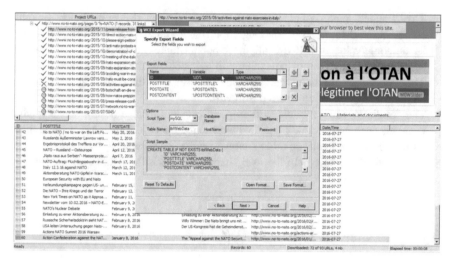

Fig. 3.10 Screenshot of web content extractor

3.10 Blogtrackers

Blogtrackers is a web application that resulted from a plethora of academic research in a variety of fields such as information science, computer science, social science, and psychology. It is developed by the Collaboratorium for Social Media and Online Behavioral Studies (COSMOS) at the University of Arkansas at Little Rock. It provides a wide variety of analysis/features to analyze blogs. Blogtrackers is a free tool that can be used by students, researchers, or any other civil or government entity. Any interested user can register and use the tool through the following URL http://blogtrackers.host.ualr.edu.

Using Blogtrackers a user can analyze the blogosphere at the *blogger level* and *blog level*. A user can search the blogosphere based on *keywords*, then track the blog posts/blogs and bloggers who mention these keywords by setting up a "Tracker." A user can check how sentiments change over time for a specific keyword or topic, check a blogger's posting activity (e.g., daily, weekly, monthly, or yearly posting frequency), explore the entities (e.g., organizations, persons, locations) related to the searched keyword(s), analyze a network of blogs or bloggers—a very unique feature as blogs do not have a natural network—based on co-occurrence, metadata (such as Web Tracker Codes (WTC)), or any shared media (e.g., shared a YouTube channel, Facebook page). A user can also submit a blog of interest to be crawled

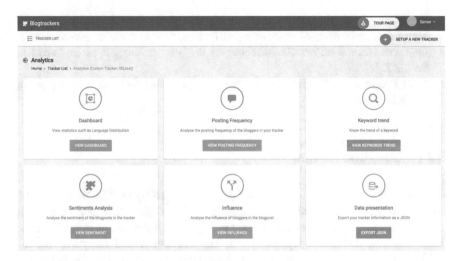

Fig. 3.11 The analytics page of Blogtrackers, showing all the current features

by Blogtrackers crawler. Then once the crawling done a user gets a notification that the data from the blog is ready to be analyzed using a variety of analytical capabilities that Blogtrackers provide. A user can also extract/export/download the data of interest from Blogtrackers in JSON format.

At the time of writing this book, Blogtrackers contains 264 blogs that have a total of 308,685 blog posts and these blog posts are written by 7410 bloggers. The blogs are hosted on servers located in 18 different countries. The blog posts are written in 46 languages. Blogtrackers database also has around 765,654 unique entities with 930 types extracted from the collected blog posts. Blogtrackers contains data that dates back to 2009.

We used Blogtrackers during many case studies such as the socio-political crisis in Venezuela [10], the European migrant crisis [9], and the anti-NATO narratives that were disseminated during many NATO exercise conducted in Europe [15]. Figure 3.11 shows the current features page that Blogtrackers has. The features are:

- *Dashboard* feature: provides high-level statistics about the blogs and bloggers included in the selected tracker, e.g., blogs language distribution, overall sentiments.
- *Posting Frequency* feature: provides information about the bloggers posting frequency/habits.
- *Keyword Trend* feature: using this feature you can track the trend of keyword(s) over a period of time.

- *Sentiments Analysis* feature: provides visualizations that show the trends of sentiments and variety of tones using various psycholinguistic measures.
- *Influence* feature: shows the influence of the blog or blogger as explained in Sect. 2.2.4.
- *Data Presentation* feature: enables the user to export the data of a tracker in JSON format.

3.11 YouTubeTracker

YouTubeTracker is a web application that is developed by COSMOS and can be used to track, monitor, and identify influential YouTube groups and content. It allows users to gain insights into content engagement behaviors of individuals via likes, dislikes, comments, replies, shares, etc. Through visual analytics, YouTubeTracker can help identify trends, opinions, communities, anomalous behaviors such as bots, Spam, Internet trolls, among other capabilities. Users can visualize networks among YouTubers, commenters, and content. YouTubeTracker is a free tool that can be used by students, researchers, or any other civil or government entity. Any interested user can register and use the tool through the following URL http://youtubetracker.host.ualr.edu.

Using YouTubeTracker a user can analyze YouTube at the *channel level* and *video level*. A user can search YouTube based on *keyword(s)*, then track the videos or channels who posted these videos by setting up a "Tracker." A user must provide his or her YouTube API key[16] to be able to collect and analyze the data.

Using YouTube Data API, YouTubeTracker can access, view, and analyze the channel name, channel creation date, channel description, related channels, location (if provided), social media footprint,[17] published date of the videos, the number of subscribers, views, likes, dislikes, comments, and the comment text for all the videos published by the channel. Additionally, for each comment, the commenter's ID, commenter's name, comment text, likes and replies on the comments, and published date can be collected.

For the video level analysis, a user can view and analyze the video id, title, posting channel id, published date, video description information, the total number of views, likes, and dislikes. The same information is collected of all the related videos for every video published by the channel. A user can also extract/export/download the data of interest from YouTubeTracker in JSON format [16–18].

---

[16]To create your YouTube API key, check out this YouTube video tutorial https://www.youtube.com/watch?v=pP4zvduVAqo.

[17]Social media footprint includes the various social media platforms that the YouTuber use to disseminate his/her videos.

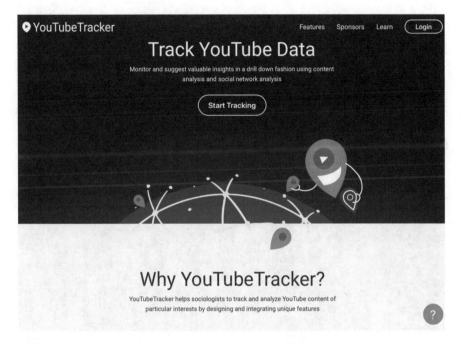

Fig. 3.12 The analytics page of Blogtrackers, showing all the current features

At the time of writing this book, YouTubeTracker database contains more than 1750 videos that are posted by more 435 YouTube channels. We used YouTube-Tracker to study various propaganda campaigns such as studying a YouTube channel that was pushing content promoting conspiracy theories. We were able to identify signals that could be used to detect such deviant content (e.g., videos, comments) and help in stemming the spread of disinformation [16]. Figure 3.12 shows the landing page of YouTubeTracker.

3.12 Botometer

Botometer is a product that is developed by Indiana University Network Science Institute and the Center for Complex Networks and Systems Research. It gives Twitter users a score between 0 and 100 based on the user activities, where a score closer to 0 means the account is more likely a human and a score closer to 100 means the account is more likely a social bot. It's available in two versions one is accessible through a web interface available at https://botometer.iuni.iu.edu/ and the other is an

Fig. 3.13 The Botometer web interface showing a Twitter bot account has a bot score of 98%

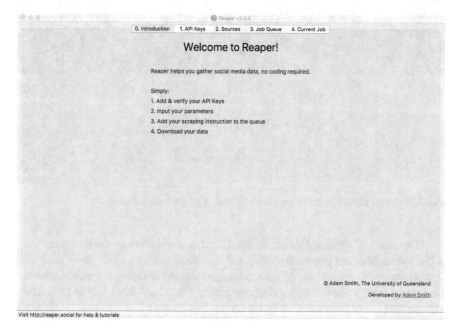

Fig. 3.14 Reaper GUI showing the welcome page

API that can be used in your own application. The API has two versions one is the "Free API" which gives 17,280 requests per day while the other is the "Pro API" which gives 43,200 requests per day and requires a paid subscription. Figure 3.13 show a Twitter bot account (@naijacupid1, currently suspended by Twitter) that was disseminating ISIL's beheading videos-based propaganda in 2015 [8].

3.13 Reaper: Social Media Scraping Tool

Reaper is a multi-platform data collection software. It is developed by the University of Queensland. Reaper is freely available and can be downloaded from https://reaper.social. It has a very simple and intuitive GUI (see Fig. 3.14) that works on both Windows and Mac OS. Reaper *v*2.5.4 can download data from:

- *Facebook* using:
  - *Facebook().page()*
  - *Facebook().post()*
  - *Facebook().comment()*
  - *Facebook().album()*
  - *Facebook().photo()*
  - *Facebook().video()*
  - *Facebook().live_video()*
  - *Facebook().user()*
  - *Facebook().status()*
  - *Facebook().event()*
  - *Facebook().group()*

- *Reddit* using:
  - *Reddit().search()*
  - *Reddit().user()*
  - *Reddit().subreddit()*
  - *Reddit().thread()*

- *YouTube* comments using:
  - *YouTube().search()*
  - *YouTube().channel()*
  - *YouTube().video()*

- *Pinterest* using:
  - *Pinterest().user()*
  - *Pinterest().board_pins()*
  - *Pinterest().pin()*

- *Twitter* using:
  - *Twitter().search()*
  - *Twitter().user()*

- *Tumblr* using:
  - *Tumblr().blog_info()*
  - *Tumblr().tag_posts()*

Reaper uses the user's API credentials to download data from social media channels. To use this tool, users need to create an *app* with any or all of the aforementioned social media channels then provide the API credentials (i.e., API key, API secret, access token, access token secret, application id, and application secret) to Reaper so it can collect data with zero coding. Reaper accepts an input file in CSV format (e.g., a list of Twitter users, Facebook accounts, or YouTube videos IDs) and will give the data in CSV format as well.

Table 3.1 Summary of the tools introduced in this chapter and their capabilities

Tool name	Data collection	Data analysis	Data visualization
TouchGraph SEO browser	✓	X	✓
TAGS & TAGSExplorer	✓	✓	✓
NodeXL	✓	✓	✓
Gephi	✓	✓	✓
CytoScape	X	✓	✓
LIWC	X	✓	X
ORA	X	✓	✓
IBM Watson Analytics	✓	✓	✓
WCE	✓	X	X
Blogtrackers	✓	✓	✓
YouTubeTracker	✓	✓	✓
Botometer	✓	X	X
Reaper	✓	X	X

Table 3.1 shows a summary of the tools introduced in this chapter and their capabilities in terms of data collection, analysis, and visualization. In the next chapter, we provide a SCF tool along with a set of methodologies that can be used to uncover hidden relations among various ODGs.

References

1. S. Al-khateeb, N. Agarwal, Analyzing deviant cyber flash mobs of ISIL on twitter, in *Social Computing, Behavioral-Cultural Modeling, and Prediction* (Springer, Cham, 2015), pp. 251–257
2. S. Al-khateeb, N. Agarwal, The rise & fall of# NoBackDoor on Twitter: the Apple vs. FBI case, in *Advances in Social Networks Analysis and Mining (ASONAM), 2016 IEEE/ACM International Conference on* (IEEE, Piscataway, 2016), pp. 833–836
3. S. M. Research Foundation. Nodexl: Network overview, discovery and exploration for excel. [Online]. Available: http://nodexl.codeplex.com/wikipage?tit
4. M.A. Smith, B. Shneiderman, N. MilicFrayling, E. Mendes Rodrigues, V. Barash, C. Dunne, T. Capone, A. Perer, E. Gleave, Analyzing (social media) networks with NodeXL, in *Proceedings of the Fourth International Conference on Communities and Technologies* (ACM, New York, 2009), pp. 255–264. [Online]. Available: http://dl.acm.org/citation.cfm?id=1556497
5. J.A. Carter, S. Maher, P.R. Neumann, #Greenbirds: measuring importance and influence in Syrian foreign fighter networks. [Online]. Available: http://bit.ly/1mdXAdW
6. T. Khaund, K.K. Bandeli, M.N. Hussain, A. Obadimu, S. Al-Khateeb, N. Agarwal, Analyzing social and communication network structures of social bots and humans, in *2018 IEEE/ACM International Conference on Advances in Social Networks Analysis and Mining (ASONAM)* (IEEE, Piscataway, 2018), pp. 794–797
7. T. Khaund, S. Al-Khateeb, S. Tokdemir, N. Agarwal, Analyzing social bots and their coordination during natural disasters, in *International Conference on Social Computing, Behavioral-Cultural Modeling and Prediction and Behavior Representation in Modeling and Simulation* (Springer, Cham, 2018), pp. 207–212

8. S. Al-khateeb, N. Agarwal, Examining botnet behaviors for propaganda dissemination: a case study of ISIL's beheading videos-based propaganda, in *2015 IEEE International Conference on Data Mining Workshop (ICDMW)* (IEEE, Piscataway, 2015), pp. 51–57

9. M. Hussain, K.K. Bandeli, M. Nooman, S. Al-khateeb, N. Agarwal, Analyzing the voices during European migrant crisis in blogosphere, in *The 2nd International Workshop on Event Analytics using Social Media Data* (2017)

10. E.L. Mead, M.N. Hussain, M. Nooman, S. Al-khateeb, N. Agarwal, Assessing situation awareness through blogosphere: a case study on Venezuelan socio-political crisis and the migrant influx, in *SOTICS 2017 : The Seventh International Conference on Social Media Technologies, Communication, and Informatics* (2017), pp. 22–29

11. Q. Yin, Q. Chen, A social network analysis platform for organizational risk analysis–ora, in *2012 Second International Conference on Intelligent System Design and Engineering Application (ISDEA)* (IEEE, Piscataway, 2012), pp. 760–763

12. S. Al-khateeb, K.J. Conlan, N. Agarwal, I. Baggili, F. Breitinger, Exploring deviant hacker networks (DHN) on social media platforms. J. Digit. Forensic Secur. Law **11**(2), pp. 7–20. [Online]. Available: http://bit.ly/2nKwNJE

13. IBM Watson Analytics - Overview - United States (2017). [Online]. Available: https://www.ibm.com/us-en/marketplace/watson-analytics

14. M.N. Hussain, A. Obadimu, K.K. Bandeli, M. Nooman, S. Al-khateeb, N. Agarwal, *A Framework for Blog Data Collection: Challenges and Opportunities* (IARIA XPS Press, Venice, 2017)

15. S. Al-khateeb, M.N. Hussain, N. Agarwal, Analyzing Deviant Socio-technical behaviors using social network analysis and cyber forensics-based methodologies, in *Big Data Analytics in Cybersecurity*, ser. Data Analytics Applications (CRC Press, New York, 2017), pp. 263–280

16. M.N. Hussain, S. Tokdemir, N. Agarwal, S. Al-khateeb, Analyzing disinformation and crowd manipulation tactics on YouTube, in *2018 IEEE/ACM International Conference on Advances in Social Networks Analysis and Mining (ASONAM)* (IEEE, Piscataway, 2018), pp. 1092–1095

17. M.N. Hussain, S. Tokdemir, S. Al-khateeb, K.K. Bandeli, N. Agarwal, Understanding digital ethnography: socio-computational analysis of trending YouTube videos, in *International Conference on Social Computing, Behavioral-Cultural Modeling & Prediction and Behavior Representation in Modeling and Simulation (SBP-BRiMS 2018)* (2018)

18. S. Tokdemir, N. Agarwal, YouTube data analytics using YouTubeTracker

Chapter 4
Social Cyber Forensics (SCF): Uncovering Hidden Relationships

Abstract In this chapter, we will introduce the concept of social cyber forensics (SCF) and its usability. Then, we will introduce a tool, i.e., Maltego that can be used to study the cross-media affiliation and uncover hidden relations among different ODGs. In Sect. 4.2 we will provide three stepwise methodologies that can be followed to reach the desired outcome of analysis (e.g., uncovering the hidden relationship between Twitter accounts and a set of websites or blog sites, websites or blog sites and other websites or blog sites, and infer the ownership of a set of websites or blog sites). These methodologies have been tested during many cyber propaganda campaigns to associate and infer the relationships between different online groups. We also provide hands-on exercises to practice each of the methodologies introduced.

Keywords Social Cyber Forensics (SCF) · Maltego · Open source information · Hands-on exercises

4.1 Social Cyber Forensics Analysis (SCF) Using Maltego

Most scientific work to date has focused on the acquisition of social data from digital devices as well as applications installed on them. Over time, online social networks (OSNs) have become the largest and fastest growing entities on the Internet, containing data for hundreds of millions of people, and now even bots. OSNs hosted on platforms like Facebook, LinkedIn, and Twitter contain a plethora of data about its members, which is of interest to both cyber forensic scientists and practitioners [1]. The forensic potential of these OSNs has been acknowledged by researchers, and there have been a number of studies on extracting this forensically relevant data from them. As OSNs continuously replace traditional means of digital storage, sharing, and communication [1], collecting this ever-growing volume of data is becoming a challenge. Within the past decade, data collected from OSNs has already played a major role as evidence in criminal cases, either as incriminating evidence or to confirm alibis [2–4].

S. Al-khateeb, N. Agarwal, *Deviance in Social Media and Social Cyber Forensics*, SpringerBriefs in Cybersecurity, https://doi.org/10.1007/978-3-030-13690-1_4

Online deviant groups (ODGs), e.g., terrorist groups and criminal organizations continue to utilize OSNs to promote, enhance, and facilitate their respective goals. It might be more efficient to take an intelligence-driven approach for identifying evidentiary trials. Harvesting forensically relevant data directly from the metadata associated with the OSN user accounts themselves, as this chapter aims to discuss, would be more efficient than traditional forensic techniques of analyzing the hardware, network traffic, file systems, and other traditional scenarios in cyber forensics.

For the last three and half decades cyber forensics tools have been evolved from simple tools, which were used mainly by law enforcement agencies to important tools for detecting and solving corporate fraud [5]. Cyber forensics tools are not a new type of tools but they are evolving over time to have more capabilities, more exposure to the audience (i.e., investigators or public users), and more types of data that can be extracted using each tool. Cyber forensics tools can be traced back to the early 1980s when these tools were mainly used by government agencies, e.g., the Royal Canadian Mounted Police (RCMP) and the US Internal Revenue Service (IRS) and were written in Assembly language or C language with limited capabilities and less popularity. As time passes these tools got more sophisticated and in the mid of 1980s these tools were able to recognize file types as well as retrieve lost or deleted files, e.g., XtreeGold and DiskEdit by Norton. In 1990s these tools become more popular and also have more capabilities, e.g., they can recover deleted files and fragments of deleted files such as Expert Witness and Encase [6]. Nowadays, many tools are available for public use and have data collection and visualization capabilities that make the process of analyzing the collected data easier, e.g., Maltego tool.

Before introducing the concept of *social cyber forensics (SCF)* let's explain what *forensics* mean. The word *forensics* refer to the discipline of collecting, analyzing, and reporting of evidence to build a case (in law) or to assert a relationship among two or more entities. *Cyber forensics* also known as *computer forensics* is the discipline of using digital tools to find digital evidence to support an assertion about a relationship [7]. Cyber forensic evidence can be used to detect or prevent a crime or address a dispute from information drawn from a digital investigation. We define *social cyber forensics (SCF)* as a branch of cyber forensics which is the process of investigating the relationships among "entities" and revealing the digital connections among them in social media space by extracting/collecting metadata associated with their social media accounts, e.g., affiliations of the user, geo-location, and IP address. An *"entity"* is an information actor which can be a single individual, a group, an organization, a nation-state, etc. An entity, for example, can be a single "individual" that own a single blog or multiple blogs.

To be able to use social cyber forensics (SCF) to uncover hidden relationships among different entities an analyst need a "seed." A *seed* is an initial knowledge that can be used to investigate an entity or set of entities, e.g., a seed can be a *web tracker code (WTC)*, a blog site, a Twitter account, IP address, or any other information that the SCF tools can use to reveal hidden or unknown information about the seed entity. Web tracker code (WTC) is an online analytics tool that allows a website owner to gather some statistics about their website visitors such as their browser, operating system, and the country they are from, along with other metadata. These web trackers have an ID number that is usually embedded in the website HTML code [8–10]. For example, Google provides users with a capability to track their website activities using Google Analytics service.[1]

Social cyber forensics analysis depends completely on publicly available sources of data, i.e., it uses *open source information (OSINF)* to get the metadata needed about an entity. There are many companies that provide services to accomplish this task, e.g., PeekYou,[2] Metagoofil,[3] and Shodan,[4] but there are very few useful tools that can obtain data from multiple sources (e.g., Shodan data is accessible through Maltego) and report them in an easy to understand format like Maltego. Figure 4.1 shows the currently available tools that can do SCF. An interactive version of this figure can be accessed at http://osintframework.com/, click on OSINT Framework → Tools, to explore the available tools.

Maltego is developed by Paterva Ltd[5] and can be used to gather any publicly available data. For example, it can be used to provide an insight on how different social media platforms such as blog sites are connected or affiliated to Twitter accounts [8]. Basically, Maltego is an open source information and forensics application that can determine the relationships and real-world links among people, groups of people (social networks), companies, organizations, websites, and Internet infrastructure, e.g., Domains, DNS names, Netblocks, and IP addresses. Maltego saves an analyst a lot of time in mining and gathering information as well as representing this extracted information in an easy to understand format. Below we are providing a set of definitions of the terminologies that you will come across if you use Maltego. These terminologies will help the readers understand (later in Sect. 4.2) the methodologies provided.

- *Transforms:* tiny pieces of code that take one type of information and change it into another for interoperability.
- *Pipeline:* a set of transforms and filters that are executed in sequence, i.e., like a macro in Microsoft Excel.
- *Trigger:* a graph condition and a transform, i.e., when something happens on the graph, then run this transform.
- *Feeder:* a mechanism to feed entities into Maltego.
- *Machine:* a combination of pipelines, triggers, and feeders.

---

[1]Google Analytics: https://analytics.google.com/.

[2]PeekYou: http://www.peekyou.com/.

[3]Metagoofil: http://www.edge-security.com/metagoofil.php.

[4]Shodan: https://www.shodan.io/.

[5]Paterva Ltd: https://www.paterva.com/.

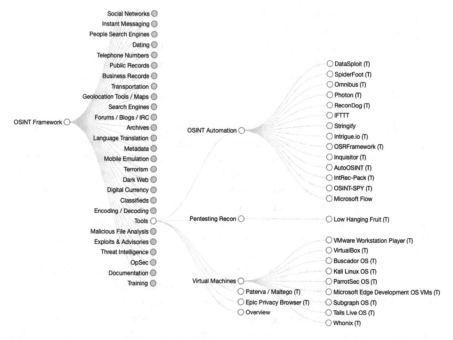

Fig. 4.1 The available OSINT tools. Not many tools available

Maltego uses Java, so it runs on any modern computer with Windows, Mac, or Linux operating systems; however, below is the recommended system requirements provided by Paterva Ltd to run Maltego:

- Windows 7 or above, Mac OS X or above, or the latest version of Linux operating system.
- Java 8.0 (or the latest version).
- At least 2GB of random-access memory (RAM), but the more the better.
- Any modern multi-core processor is adequate.
- 4GB of disk space is adequate.
- Mouse to make navigating the graphs much easier.
- Internet access is a must to fully operate. The outgoing connections will be on the following ports: 80, 443, 8081, and port 5222.

Next, in Sect. 4.2 we provide a set of stepwise methodologies that can be followed using Maltego to reach the desired outcome of analysis. Then, to practice these methodologies we provide hands-on exercises in Sect. 4.3.

4.2   Methodologies to Extract Open Source Information

In this section, we provide three sets of stepwise methodologies that can be followed to answer some interesting questions like the below ones.

1. Given a set of websites or blog sites (seed), can we *discover* other unknown websites that are working together with the "seed" to disseminate propaganda or misinformation?
2. Given a set of Twitter accounts or other social media accounts can we find the footprint of the seed accounts on other social media channels, i.e., their cross-media affiliation?

   - Can we identify other accounts working with the seed to disseminate propaganda?

3. Given a set of websites or blog sites (seed), can we *infer the ownership* of these websites, i.e., who they belong to individuals or organizations?

   - Are these set of websites working together to disseminate propaganda?
   - What are their strategies?

These methodologies are tested on many real-world events such as the NATO Trident Juncture 2015 exercise (TRJE2015),[6] Exercise Anakonda 2016 (AN16),[7] and Brilliant Jump 2016 exercise (BRJP16).[8] In all the aforementioned events these methodologies proved to be effective in uncovering hidden connections between different ODGs, or show their cross-media affiliation which gave great insights to many Public Affair Officers (PAO). The methodologies provided next are tested and used with the free version of Maltego, i.e., *Maltego CE v4.0.11*.

4.2.1   Finding Related Websites From Web Tracker Code (WTC)

Here, we introduce the first stepwise methodology that can be followed to answer the first question mentioned earlier in Sect. 4.2. A hands-on exercise to practice this methodology is provided later in Sect. 4.3.1. This methodology has the following steps:

1. Insert the seed blogs (all must start with "www.") into Maltego by choosing "*Infrastructure* → *Website*" then drag and drop the entity to the main window.

---

[6]Trident Juncture 2015 exercise: https://jfcbs.nato.int/trident-juncture.

[7]Exercise Anakonda 2016: http://www.eur.army.mil/anakonda/.

[8]Brilliant Jump 2016 exercise: https://tinyurl.com/y7we5q34.

2. Select all the inserted blog sites and right-click, then use "*ToTrackingCodes*" transformation to get the Unique Identifiers for these seed blogs, e.g., Google Analytics IDs.[9]
3. Select all the tracker codes and right-click, then use "*ToOtherSitesWithSame Code*" transformation to get the blogs that use the same code.[10]
4. Repeat steps 2 and 3 until no new blogs or tracking codes are identified.
5. Select all the blog sites and choose "*ToIPAddress[DNS]*" transformation to get the IP addresses of all the sites.
6. Select all the identified IP addresses and choose "*ToLocationCountry*" transformation to get the location of where these websites geolocated.

4.2.2 Finding Blog Sites From Twitter Handles

Here, we introduce the second stepwise methodology that can be followed to answer the second question mentioned earlier in Sect. 4.2. A hands-on exercise to practice this methodology is provided later in Sect. 4.3.2. This methodology has the following steps:

1. Copy and paste the Twitter username (without the @ sign) to the main window.
2. Right-click on the Twitter username, then change the type to "*People → UnknownSuspect*."
3. Right-click on the Unknown Suspect entity and use "To Twitter Affiliation [Search Twitter]" transformation to find all possible accounts this person have.
4. Select the Twitter username and use "*ToPerson[Convert]*" transformation.
5. Select the person entity and use "*ToEmailAddress[UsingSearchEngine]*" transformation to find possible email addresses for that person using a search engine.
6. Select the email entity and use "*ToWebsite[usingSearchEngine]*" transformation to find websites associated with the email address found.
7. Select the email entity and use "*ToURLs[Showsearchengineresults]*" transformation to find any URLs associated with the email address found.
8. If steps 6 and 7 did not return good or enough results, then: Select the person entity and use "*ToWebsite[usingSearchEngine]*".
9. Note that the **methodology in Sect. 4.2.1** can be applied here to all the websites discovered in **steps 6 and 7** to discover more blogs.

---

[9]If the website points to a WTC, it means that the website HTML code contains the WTC.

[10]If the WTC point to the website, it does not necessarily mean that the WTC is embedded in the website HTML, but it does mean that the WTC owner added that website to their list of to be tracked websites.

4.2.3   Inferring the Ownership or Hidden Connections Among Different Websites

Here, we introduce the third stepwise methodology that can be followed to answer the third question mentioned earlier in Sect. 4.2. A hands-on exercise to practice this methodology is provided later in Sect. 4.3.3. This methodology has the following steps:

1. Insert the seed websites (all must start with "www.") into Maltego by choosing "*Infrastructure → Website.*"
2. Select all the inserted blog sites and right-click, then use "*ToTrackingCodes*" transformation to get the Unique Identifiers for these seed blogs, e.g., Google Analytics IDs.
3. Select all websites and use "*ToIPAddress[DNS]*" transformation to get the IP addresses of all the sites.
4. Select all IP Address and use "*ToEntitiesfromWhois[Alchemy]*" transformation to get the owner name, location, phone number, email address, etc.
5. Select all IP Address again and use "*toLocation[country]*" transformation.
6. Select all IP Address and use "*ToTelephoneNumber[fromwhoisInfo]*" transformation.
7. Select all IP Address and use "*ToEmailAddress[fromwhoisInfo]*" transformation.
8. Note that the **methodology in Sect. 4.2.1** can be applied here to the inserted blog sites **step 1** to discover more blogs if needed.

4.3   Hands-On Exercises

In this section, we provide hands-on exercises to practice the aforementioned methodologies (Sect. 4.2). Note that due to the nature of these websites and Twitter accounts (i.e., deviant nature), they might be taken down or suspended by the time the reader attempts these exercises. If this happens, then feel free to use any blog site or Twitter account of your interest. To practice the aforementioned methodologies download Maltego from their website, register for an account, then once you launch Maltego CE login with your credentials. After logging in, create a new graph and follow the specified methodology along with the "seed" knowledge provided to start your practice.

4.3.1   Exercise A

For this exercise, consider the blog site: www.globalresearch.ca as your seed knowledge. Your task is to find all the other websites that are connected to it. To do so, you will need to follow all the steps in Sect. 4.2.1. Once you do so, your result should look similar to Fig. 4.2. Figure 4.2 shows that the given seed website has two WTCs which are connected to 18 websites through different types of relations.

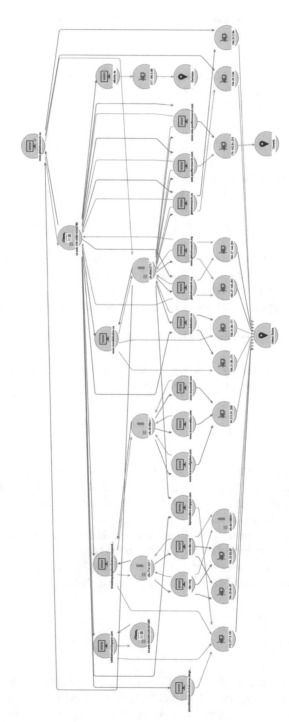

Fig. 4.2 Maltego results after following the steps in Sect. 4.2.1

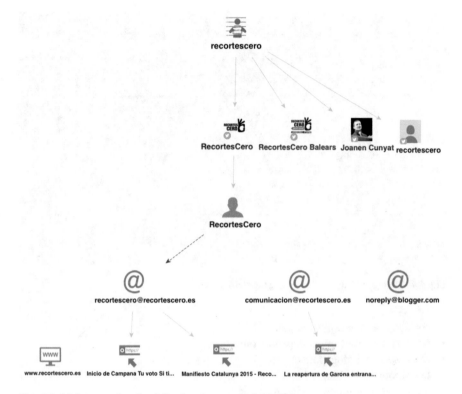

Fig. 4.3 Maltego results after following the steps in Sect. 4.2.2

4.3.2   Exercise B

For this exercise, consider the Twitter account: @recortescero as your seed knowledge. Your task is to find all the blog sites that are connected to this Twitter account. To do so, you will need to follow all the steps in Sect. 4.2.2. Once you do so, your result should look similar to Fig. 4.3. Figure 4.3 shows that the given seed Twitter account has three email addresses which are related/connected to four websites.

4.3.3   Exercise C

For this exercise, consider the set of websites provided below as your seed knowledge:

- www.zygosoccerreport.com
- www.zurumnewsdigest.blogspot.com
- www.zurumnewsdigest.blogspot.co.uk

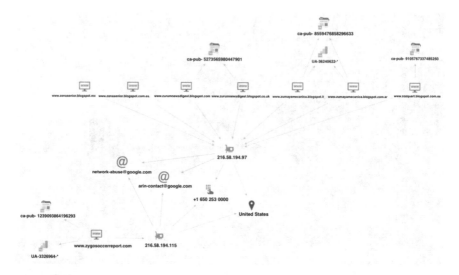

Fig. 4.4 Maltego results after following the steps in Sect. 4.2.3

- www.zumayamecanica.blogspot.it
- www.zumayamecanica.blogspot.com.ar
- www.zozquert.blogspot.com.es
- www.zonasenior.blogspot.mx
- www.zonasenior.blogspot.com.es

Your task is to uncover all the hidden relationships among them. If you follow all the steps in Sect. 4.2.3 you should obtain the results as shown in Fig. 4.4. By analyzing the graph you should find out that all these seed websites are connected by two email addresses, one phone number, and share the same geo-location.

In the next chapter, we will provide a brief overview of various case studies that leveraged the tools and methodologies introduced in this book so far. So, basically next chapter will bring all the concepts, tools, and methodologies covered so far together.

References

1. M. Huber, M. Mulazzani, M. Leithner, S. Schrittwieser, G. Wondracek, E. Weippl, *Social Snapshots: Digital Forensics for Online Social Networks* (ACSAC, Orlando, 2011), pp. 113–122
2. V. Juarez, *Facebook Status Update Provides Alibi*. Available: http://cnn.it/2mUOo48
3. E. Grube, Assault Fugitive who was Found via Facebook is Back in NY. Available: http://newyorkcriminallawyersblog.com/2010/03/assault-criminal-who-was-found-via-facebook-is-back-in-ny.html

4. M. Fisher, Facebook: A Place to Meet, Gossip, Share Photos of Stolen Goods. Available: http://www.washingtonpost.com/wp-dyn/content/article/2010/12/14/AR2010121407423.html
5. N. Alherbawi, Z. Shukur, R. Sulaiman, Systematic literature review on data carving in digital forensic. Procedia Technol. **11**, 86–92 (2013)
6. K. Oyeusi, *Computer Forensics*. Available: http://docslide.us/documents/computer-forensics-558454651e7df.html
7. D. Povar, V. Bhadran, Forensic data carving, in *Digital Forensics and Cyber Crime*. Lecture Notes of the Institute for Computer Sciences, Social Informatics and Telecommunications Engineering, vol. 53 (Springer, Berlin, 2011), pp. 137–148. Available: http://bit.ly/2mzILFW
8. L. Alexander, *Open-Source Information Reveals Pro-Kremlin Web Campaign*. Available: https://globalvoices.org/2015/07/13/open-source-information-reveals-pro-kremlin-web-campaign/
9. M. Bazzell, *Open Source Intelligence Techniques: Resources for Searching and Analyzing Online Information*, 4th edn. (CCI, Charleston, 2014). Available: https://inteltechniques.com/book1.html
10. S. Al-khateeb, M.N. Hussain, N. Agarwal, Social cyber forensics approach to study twitter's and blogs' influence on propaganda campaigns, in *International Conference on Social Computing, Behavioral-Cultural Modeling and Prediction and Behavior Representation in Modeling and Simulation* (Springer, Berlin, 2017), pp. 108–113

Chapter 5
Case Studies of Deviance in Social Media

Abstract In this chapter, we provide a high-level view of various case studies that include *deviant acts*, *deviant events*, and *deviant groups*. This chapter utilizes the concepts, tools, and methodologies that we presented throughout this book to study, analyze, and have a better understanding of many real-world deviant events, acts, and groups. Here, we briefly highlight case studies of the work we conducted during many deviant cyber campaigns that were projected against NATO forces. In addition to these cases, we shed light on the deviant cyber campaigns conducted by the so-called Islamic State, also known as ISIL, ISIS, or Daesh. We will highlight the major findings of these cases and point the interested reader for more details to other published literature.

Keywords Deviance in social media · Case studies · ISIS · ISIL · Daesh · Anti-NATO propaganda · Dragoon ride exercise · Crimean water crisis · Beheading propaganda

5.1 Introduction

In this section, we introduce the overall methodology that we followed to study *deviant acts* such as cyber propaganda, misinformation (i.e., misleading), and disinformation (i.e., lies) campaigns that are carried on by *deviant groups* during various events on different social media channels, e.g., Twitter, blogs, and YouTube. The overall methodology is illustrated in Fig. 5.1. This methodology has been tested on many case studies, including the four cases highlighted below and provided consistent results in all of the cases.

The methodology can be followed whenever a deviant event occurs. The first step of the methodology involves identifying related keywords to that event by domain experts. Then, we search for platform-specific keywords using TouchGraph SEO Browser (highlighted in Sect. 3.1) to identify the most used keywords in a specific social media, e.g., Twitter. Third, we search various online social media platforms to identify an initial seed of data, e.g., Twitter accounts tweeting propaganda about the event, or a YouTube video containing propaganda (e.g., tweets contain an image

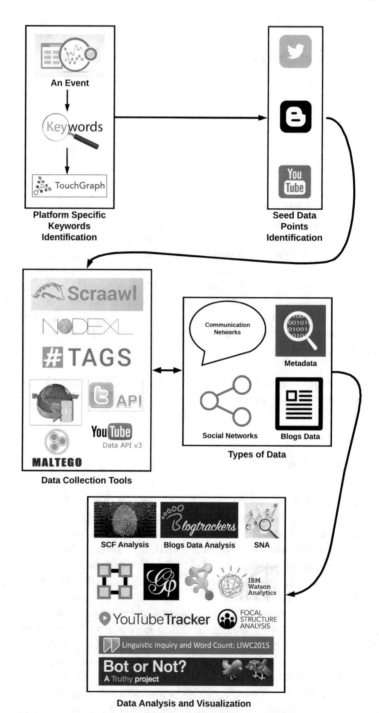

Fig. 5.1 A high-level view of the methodology followed to study various deviant acts and groups

Fig. 5.2 Examples of offensive and biased memes observed on blogosphere during NATO's exercises to delegitimize exercises' objectives

or a URL to blog post with offensive or biased memes, see Fig. 5.2), or a YouTube video that contains specific narratives such as conspiracy theory videos. Fourth, using various data collection tools (e.g., NodeXL, Scraawl, Web Crawlers (e.g., Web Content Extractor), YouTube API, Twitter API, TAGS, and Maltego) we extract the social and communication networks of Twitter users, crawl blog data, identify other social media footprint, identify bots and bots coordination strategies, and extract the metadata associated with these deviant groups accounts. Finally, we conduct a set of analyses on all the data and derive insights. These analyses include:

- Social Cyber Forensics (SCF) Analysis: to identify relations among various groups, uncover their cross-media affiliation, and identify more deviant groups using the methodologies explained in Chap. 3.
- Blogs Data Analysis: to identify leading narratives, influential blogs and bloggers, sentiment analysis, keywords trends, etc., using Blogtrackers tool (highlighted in Sect. 3.10).
- Social Network Analysis (SNA): to identify the creators of the narratives and identify the role of various nodes in the network, e.g., the source of information/top disseminators (i.e., nodes with high outdegree centrality), brokers (i.e., nodes with high betweenness centrality), and type of nodes (e.g., bot or human accounts).

The aforementioned analysis is conducted using the various data analysis and visualization tools introduced in this book such as Gephi (highlighted in Sect. 3.4), CytoScape (highlighted in Sect. 3.5), LIWC (highlighted in Sect. 3.6), ORA (highlighted in Sect. 3.7), IBM Watson Analytics (highlighted in Sect. 3.8), Blogtrackers (highlighted in Sect. 3.10), YouTubeTracker (highlighted in Sect. 3.11), and Botometer (highlighted in Sect. 3.12). Next, we introduce four case studies and highlight the main points for each. The details of each case study are published and a reference to each case is provided for interested readers.

5.2   Case Study 1: Propaganda During the 2014 Crimean Water Crisis

Russia's annexation of the Crimean peninsula on March 16, 2014, met with international discontent. Both the United Nations (the UN) and the NATO Secretary General, Anders Fogh Rasmussen, have condemned this expansion of the Russian sphere of influence. Civil unrest and political instabilities in both Russian-annexed Crimea and Ukraine resulted in significant humanitarian crises due to economic impacts, changes in civil authority, and deep uncertainties about shifting political and economic relationships. Grievances, requests for help, and on-the-ground reports on the developing conflict were reported on a variety of open source platforms including blogs, news websites, Twitter, Facebook, and other open source channels such as YouTube.

The economic impact of the annexation dominated online media coverage. Several stories published by Russian news agencies, including Telegraph Agency of the Soviet Union (TASS) or Information Telegraph Agency of Russia (ITAR), claimed that Ukraine's government had ceased work on the North Crimean Canal that carries water from the Dnepr river to Crimea [1]. The Russian international television network Russia Today (RT) reported that satellite images showed Ukraine deliberately trying to cut off the Crimean peninsula's water supply by building a dam, while Russian scientists were trying to find ways to supply Crimea with fresh water in the meantime [2]. A *New York Times* article reported that quality of life was deteriorating in Russian-annexed Crimea—a water shortage was observed, Crimean farms were drying, food supplies were inadequate, and price of basic goods, such as milk and gas, had doubled [3]. The article further stated that the tourism economy was also suffering and was down by one third from the previous year; few banks were operating—Ukrainian banks had closed, Russian banks were barely open, and Western banks feared sanctions for continuing to operate in Crimea; only Russian channels were providing television and cable services; and telecommunications were erratic as carriers shifted from Ukrainian to Russian providers.

The Russian media largely blamed Ukraine government officials for these problems. Several social media outlets, including blogs, picked up the pro-Russian narrative and amplified it further suggesting that Ukraine was colluding with the West in direct conflict with Russia against the will of Crimean citizens [4]. The propaganda from pro-Russian mainstream media and social media sources were further intensified by deviant social bots on Twitter. Bots effectively disseminated thousands of messages in relation to the Crimean water crisis. These bots were disseminating anti-West and pro-Russia news articles in a bid to provoke hysteria. Numerous bots were simply tweeting the same article after copying it to various websites and blogs, making it appear as if the article were independently posted on different URLs. In other words, bots were cloning the misinformation, creating an echo chamber, and misleading the public.

We used an integrated data collection strategy from disparate publicly available online sources that were identified as relevant for the crises. Often content, e.g.,

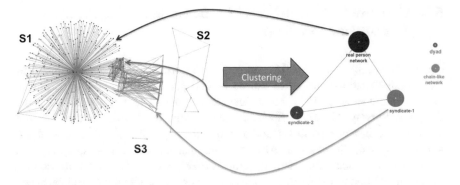

Fig. 5.3 Social bot network

reports, images, videos, and articles, originated on one social media site and was diffused on many other sites without attribution. It was therefore imperative to track multiple social media sites to identify implicit interconnections. Using hyperlinks, a snowball data collection[1] approach was used. We identified the popular blog posts for the Ukraine-Russia conflict and, by cross-referencing with Twitter data, we found which posts were diffused most often on Twitter. We collected the social (friend and follower relations) and communication (tweet, mention, and reply relations) networks of these set of bots or botnets using TweetTracker[2] [5] and NodeXL [6].

Figure 5.3 shows the social network of the identified network of bots, i.e., the botnets. This network has three sub-networks, labeled as S1, S2, and S3. Applying the Girvan–Newman clustering algorithm–an algorithm that detects communities in a network based on how closely the nodes are connected—resulted in five clusters. On the left are the expanded clusters and on the right is the collapsed view. The S1 sub-network has unusual structural characteristics. The bots that were identified in this case study are considered as "simple" because Syndicate-1 and Syndicate-2 sub-networks have simple bots that are following each other, i.e., IFYFM and FMIFY behaviors are observed in these bots (hierarchical bot structures are more complex). The details and the results of this case study were published in Al-khateeb et al. [7]. Interested readers are encouraged to read.

---

[1]Snowball sampling is a non probability sampling technique in which the sample grows like a rolling snowball, for example, in this study we started with the following keywords "Ukraine," "Ukraine Crisis," "Euromaidan," "Automaidan," and "Ukraine's Automaidan Protestors" to collect data about the crisis. Initially the dictionary of keywords for crises events is manually seeded, but evolved automatically.

[2]TweetTracker is a tool developed by Arizona State University to collect and analyze Twitter data, available at: http://tweettracker.fulton.asu.edu.

5.3   Case Study 2: Anti-NATO Propaganda During the 2015 Trident Juncture Exercise

On November 4, 2015, the US soldiers along with soldiers from more than thirty partner nations and allies moved 36,000 personnel across Europe during the Trident Juncture 2015 exercise. The exercise took place in the Netherlands, Belgium, Norway, Germany, Spain, Portugal, Italy, the Mediterranean Sea, the Atlantic Ocean, and also in Canada to prove the capability and readiness of the alliance on land, air, and sea. Also, to show that the alliance is equipped with the appropriate capabilities and capacities to face any present or future security issues. In addition to the partner nations and allies, more than twelve aid agencies, international organizations, and non-governmental organizations participated in the exercise to demonstrate "NATO's commitment and contribution to a comprehensive approach" [8].

Many opponent groups launched deviant cyber campaigns on Twitter, blogs, Facebook, and other social media platforms that encouraged citizens to protest against the exercise or conduct violent/deviant acts. We identified six deviant groups by searching for their names on various social media platforms to identify their Twitter and blogging profiles. NATO's public affairs officers then verified these profiles (i.e., the domain experts). These six groups propagate their messages on social media inviting people to act against NATO and TRJE 2015 exercise.

We identified an initial set of twelve blog sites along with the Twitter accounts (9 Twitter accounts) that are used to steer the audience from Twitter to blogs. We used Twitter API and NodeXL to collect a network of replies, mentions, tweets, friends, and followers for all the nine Twitter accounts and whoever is connected to them with any one of the aforementioned relationships. The Twitter handles, blogs, and names of the groups studied in this research are publicly available. However, in order to ensure their privacy, we do not disclose them here.

We did metadata extraction, e.g., web tracker code (WTC) using Maltego. There are many types of WTCs, for example, Google Analytics ID which is an online analytics service provided by Google that allows a website owner to gather statistics about their website visitors such as their browser, operating system, and country among other metadata. Multiple sites can be managed under a single Google Analytics account. The account has a unique identifying "UA" number, which is usually embedded in the website's HTML code[3] [9]. Using this identifier other blog sites that are managed under the same UA number can be identified. This method was reported in [9, 10].

So, using Maltego we inferred the connections among blog sites and identified new sites that were previously undiscovered. We used Maltego in a snowball manner

---

[3]If the website points to a WTC, it means that the website HTML code contains the WTC. On the other hand, if the WTC points to the website, it does not necessarily mean that the WTC is embedded in the website HTML code, but it does mean that the WTC owner added that website to their list of to be tracked websites.

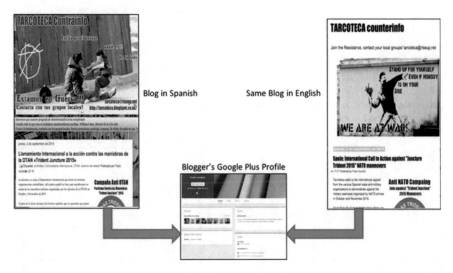

Fig. 5.4 A bridge blogger, Spanish and English blogs with anti-NATO narratives. The two blogs are kept unlinked. Connection between the blogs is discovered through Google Plus profile of the blogger using SCF methodologies

to discover other blog sites. We were able to identify additional 9 blogs that are connected to the initial seed blogs by the same Google Analytics IDs. These newly identified websites have the same content published on different portals and sometimes in different languages. For example, a website written in English may also have another identical version but written in another language (see Fig. 5.4) that is native to the region. Such blogs are also known as "bridge blogs" [11]. We collected the IP addresses, website owner name, email address, phone numbers, and locations of all the websites to infer the hidden connections among these websites. Based on the websites geo-location (estimated by their IP address geo-location) we obtained three clusters of websites (see Fig. 5.5). These clusters are helpful to know the originality of the blog sites, which would help an analyst understand the propaganda that is being pushed by the specific blog site. From an initial set of 12 blog sites, we grew to 21 blog sites, 6 locations, and 15 IP addresses. All the blog sites we identified during this study were crawled and their data is stored in a database that the Blogtrackers tool can access and analyze.

We also applied SNA to find the most important nodes in the network by activity type. Using NodeXL we were able to find the most used hashtags during the time of the exercise. This helps in targeting the same audience if counter-narratives were necessary to be pushed to the same audience. In addition to that, we found the most tweeted URLs in the graph. This gives an idea about the public opinion concerns. Finally, we found the most used domains, which helps to know where the focus of analysis should be directed, or what other media platforms are used. Additionally, we applied FSA on both the social network and the communication network to identify powerful groups of individuals that are effecting the cyber propaganda

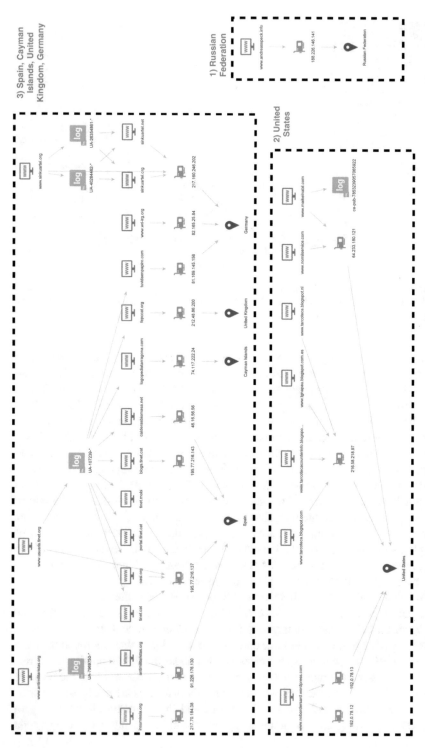

Fig. 5.5 The additional blog sites that were identified using Maltego had three clusters labeled as 1, 2, and 3 based on their IP address geo-location

campaign. Finally, we analyzed the blog data that we crawled. Using SCF analysis and SNA we were able to identify a total of 21 blog sites of interest. We trained web crawlers to collect data from these blogs, store the data in Blogtrackers database, and perform the following analyses: posting frequency, influential blogs and bloggers analysis, keyword trends, sentiment analysis, etc. In this case study, the deviant groups also used social bots to further disseminate their agenda to a large audience in a short period of time. For interested readers, the details of this study are published in [12] and an extended version of the study along with findings of the bots used during TRJE 2015 is published in [13].

5.4   Case Study 3: Anti-NATO Propaganda During the 2015 Dragoon Ride Exercise

On March 21, 2015, US soldiers assigned to the 3rd Squadron, 2nd Cavalry Regiment in Estonia, Latvia, Lithuania, and Poland as part of Operation Atlantic Resolve began Operation Dragoon Ride. The US troops, nicknamed "Dragoons," were sent on a transfer mission crossing five international borders and covering more than 1100 miles to exercise the units' maintenance and leadership capabilities, and to demonstrate the freedom of movement that exists within NATO [14].

Many opponent groups launched campaigns to protest the exercise, e.g., "Tanks No Thanks" [15], which appeared on Facebook and other social media platforms, promising large and numerous demonstrations against the US convoy [16]. Czech President Milos Zeman expressed sympathy with Russia; his statements were echoed in the pro-Russian English language media and the Kremlin-financed media, i.e., Sputnik news [17]. The website of Russia Today (RT.com) also reported that the Czechs were not happy with the procession of the "U.S. Army hardware" [15]. However, thousands of people from the Czech Republic welcomed the US convoy as it passed through their towns, waving US and NATO flags, while the protesters were not seen.

During that time many deviant social bots or impersonators bots were disseminating propaganda, asking people to protest and conduct violent acts against the US convoy. We identified a group of these bots (around 90 Twitter account) using Scraawl.[4] We collected the social (friend and follower relations) and communication (tweet, mention, and reply relations) networks of these set of bots or botnets using NodeXL.

We analyzed the friends and followers network (i.e., the social network) of these bot accounts. The social network had two sub-networks, namely S1 and S2 as shown in Fig. 5.6. The small sub-network, i.e., S2, contains only three nodes (i.e., a triad), hence it was rejected from further analysis, as it did not contribute much to the information diffusion. Since S1 is the largest sub-network, containing the majority of nodes, we examined this sub-network further.

---

[4]Scraawl is an online social media analysis tool, available at www.scraawl.com.

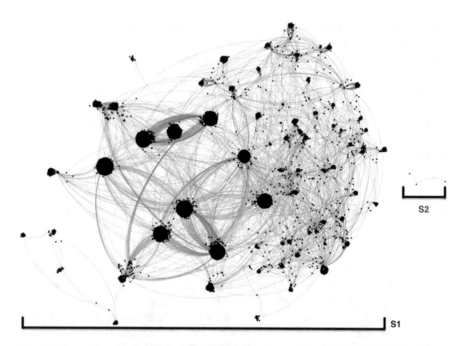

Fig. 5.6 Two sub-networks, S1 and S2. S1 and S2 are un-collapsed. Edges in green denote mutually reciprocal relations (bidirectional edges) while edges in red color denote non-reciprocal relations (unidirectional edges). Nodes are sized based on their indegree centrality

Closer examination of the S1 sub-network revealed that the members of that network were more akin to a network we call as "syndicate" network, i.e., a network that has dense connections among its members and does not have a central node, i.e., no organizational hierarchy. Further examination of the within-group ties revealed a mutually reciprocated relationship (the nodes followed each other), suggesting that the principles of "Follow Me and I Follow You" (FMIFY) and "I Follow You, Follow Me" (IFYFM)—a well-known practice used by Twitter spammers for "link farming," or quickly gaining followers [18] were in practice. This behavior was also observed during the Crimean water crisis botnet case study (highlighted in Sect. 5.2). Although there might not be a single most influential node, a group of bots may be coordinating to make an influential group. To study this behavior further, we applied the focal structures analysis (FSA) (highlighted in Sect. 2.2.5) approach to find influential group of bots [19].

In this case study, deviant groups used a sophisticated tool to disseminate their propaganda and speed up the dissemination process by using botnets. These botnets were very sophisticated compared to the Crimean water crisis in 2014. The network structure of the botnets observed in 2015 Dragoon Ride Exercise is much more complex than in 2014 case of the Crimean water crisis. Botnets in the Dragoon Ride 2015 Exercise case required a more sophisticated approach to identify the organizers or seeders of information, i.e., it required applying FSA to both the social network (friends and followers network) and the communication network (tweets, replies, and mentions network). The evolution of complexity in the bots' network structures confirms the need for a systematic study of botnet behavior to develop advanced approaches and tools that can deal with predictive modeling of botnets. The details of the study comparing the botnets used during the Crimean water crises and the Dragoon Ride Exercise are published in [20]. Interested readers are encouraged to read it.

5.5   Case Study 4: ISIS Beheading Propaganda in 2015

Studies have shown that public opinion is going to be more effective than bullets and bombs in the future war. Countering ODGs mentality is a must to win the new battle of ideas [21]. Transnational terrorist groups such as ISIL know that opinions can be influenced and they are using very sophisticated techniques to collapse the time and space by the speed of the Internet. A study conducted by the Defense Academy of the United Kingdom [21] examines the practice of sharing beheading videos of hostages by Al-Qaeda as an instance of strategic communication, defined as: "A systematic series of sustained and coherent activities, conducted across strategic, operational and tactical levels, that enables understanding of target audiences, identify effective conduits, and develops and promotes ideas and opinions through those conduits to promote and sustain particular types of behaviour" [21].

In 2015, the so-called Islamic State or Daesh started releasing videos on social media where they depicted gruesome beheading of innocent people and accusing them of being unbelievers or traitors in an attempt to spread horror among the minority religion in the Middle East and capture the attention of media. ISIL had released three videos of beheading of innocent people namely: the beheading of Egyptian Copts in Libya [22], the beheading of the Arab-Israeli "Spy" in Syria [23], and the beheading of Ethiopian Christians in Libya [24] (see Fig. 5.7). These events were almost a month apart viz. on February 15, 2015, March 10, 2015, and April 19, 2015, respectively.

Fig. 5.7 Screen capture of (**a**) the beheading of Egyptians Copts in Libya by ISIL, (**b**) the beheading of the Arab-Israeli "Spy" in Syria by ISIL, and (**c**) the beheading of the Ethiopian Christians in Libya by ISIL

We have used our methodology that is depicted in Fig. 5.1 to collect and analyze the datasets we obtained from Twitter related to the aforementioned beheading video-based propaganda of ISIL. The datasets were collected using Twitter Archiving Google Sheet (TAGS) based on specific keywords such as #beheadingChristians, #beheadingofCoptic, #younglions, Mohammed Merah, #infiltrator, and #Traitor. We collected the communication network, i.e., tweets, retweets, replies, and mentions network of the Twitter users who used these keywords or hashtags in their tweets. In addition to that, the dataset we collected contains other metadata such as the number of friends and followers of a user, geo-locations of users, the time of tweets, and the language of the user.

During these three events ISIL or Daesh has used social media in a very sophisticated way (i.e., by using social bots) to disseminate their propaganda (e.g., messages contain URLs to an image, YouTube video, or news article about the beheading) to a large audience in a very short period of time. The bots or automated social actors (ASAs) used in this case is different than the Crimean water crises (explained in Sect. 5.2) and the Dragoon Ride 2015 Exercise bots (explained in Sect. 5.4) [25]. These bots were similar to another case study conducted by Abokhodair et al. [26] during the 2012 Syrian civil war conflict. The bots, in this case, were of type generator and peripheral bots, where some accounts would tweet the propaganda while others would retweet it. These accounts are not connected through their social network but they are connected through their communication network. They also used some of the very effective information maneuvers to disseminate their messages. Below we are highlighting a few:

- Misdirection: where a bot would tweet unrelated news that is happening somewhere else but mention a hashtag related to the beheading crises.
- Hashtag Latching: when strategically associating unrelated but popular or trending hashtags to target a broader, or in some cases a very specific audience (e.g., using the #WorldCup then include a URL of beheading video).
- Smoke Screening: when a bot would mention something about ISIL but not necessarily related to the beheading. Similar techniques have been used in the Syrian Social Bot (SSB) to raise awareness of the Syrian civil war [26].
- Thread-Jacking: the change of topic in a "thread" of discussion in an open forum (e.g., using a hashtag of #ISIL but the tweet has a URL to an online shopping website).

In addition to the aforementioned information maneuvers, we find that once the media mentions an event, the popularity of the event on social media (i.e., the number of people talking about it on social media with their friends and followers, tweeting or retweeting) increases and gets highest for the first 2 days, after that the event's popularity declines. For a complete list of keywords, findings, and other details of the study, interested readers are encouraged to read [25].

Throughout this book, you were introduced to two main concepts, namely *deviance in social media* and *social cyber forensics*. After reading Chap. 1, you should be knowledgeable about the various types of *deviance in social media* such as *deviant groups*, *deviant events*, and *deviant tactics*. You should also be knowledgeable about the power of combining multiple disciplines to have a better understanding of a specific event or phenomenon. In Chap. 2 of the book, we introduced various concepts of *graph theory* and *social network analysis* to have a better understanding of what a graph means and how you can leverage various network measures to assess *central nodes* or *groups* in a graph. These concepts represent the fundamental knowledge needed to understand a network of individuals whether they are deviant or not. In Chap. 3, we introduce a set of tools that help you apply and estimate the concepts learned in Chap. 2. Chapter 4 introduces the second main concept of the book, i.e., *social cyber forensics*. We give a brief history of the evolution of social cyber forensics and demonstrate its advantages in connecting the seemingly disconnected entities. We also introduce a cyber forensic tool, viz., *Maltego*, and a set of methodologies that can be used to conduct analysis. Finally, in Chap. 5 we show a framework that brings together all the concepts, tools, and analysis we have discussed throughout the book. Further, we exemplify this framework via a set of case studies to study various real-world events.

With all the knowledge gained by reading this book, you should be able to study various events that happen in social media by collecting relevant data, analyze it, visualize your results, and derive actionable insights. All this can help in understanding various events and develop predictive models and tools that can forecast occurrence and outcomes of events.

This field, as you may have imagined by reading the book, has many challenges but also opportunities. Below we are including some of the challenges and opportunities:

- Identifying the right seed (e.g., keywords, hashtags, individuals, organizations, social media channels, etc.) data that can be used to snowball more data about an event or phenomenon is challenging, hence in addition to the *domain experts* help that can be leveraged, there is a need to build more tools that can identify top entities based on their usage on various social media platforms to identify the right seed entities and social media channels to be used to study a specific event such as *TouchGraph tool* (explained in Sect. 3.1).
- Not all social media channels provide free or unrestricted access to their data via API's (e.g., Snapchat) or even has an API (e.g., blogs). Hence there is a need to build tools that can collect data even if the social media platform does not provide an API such as *web content extractor* and *Blogtrackers* (explained in Sects. 3.9 and 3.10, respectively).
- Many social media channels do not provide unlimited access to the data that they possess. It only provides a portion, e.g., Twitter allows access to roughly 1% of the data through its REST APIs. Also, data archives can be collected up to a specific period of time, for example, Twitter allows developers to collect data that are dated back up to 7 days using its REST APIs. Hence, being aware of the events and the social media channels in advance is essential, which is an unreasonable expectation. Therefore, there is a need for more data allowance from social media platforms as well as initiatives to create open data platforms. Some efforts (with limitations) in this direction exist, including,

- The social computing data repository by Arizona State University, http://socialcomputing.asu.edu/pages/datasets
- The network data repository by the University of California at Irvine, http://networkdata.ics.uci.edu
- The data world repository, https://data.world
- Kaggel Inc. data repository, https://www.kaggle.com/datasets
- The Stanford large network dataset collection, https://snap.stanford.edu/data/
- The complex networks data repository, https://icon.colorado.edu/#!/networks
- The Koblenz Network Collection by the Institute of Web Science and Technologies at the University of Koblenz-Landau, http://konect.uni-koblenz.de/networks/
- Google Dataset Search, https://toolbox.google.com/datasetsearch

- Social media provides a huge amount of data, with a variety of data types (text, images, videos, audio, XML/JSON, etc.), and at a very rapid pace, i.e., volume, variety, and velocity—typical challenges of big data. Moreover, the data contains a lot of attributes, so identifying the right attributes is a challenge (data size and speed). Hence, data mining and machine learning techniques for dimensionality reduction can be leveraged for feature selection, extraction, and engineering. Building tools that tackle challenges of preprocessing big data is of huge help, e.g., *IBM Watson Analytics* (explained in Sect. 3.8).
- Connecting data points is a challenge when collecting data from multiple sources. Hence, there is a need to develop methodologies and tools that can be used to connect entities across different platforms, e.g., using metadata extracted by social cyber forensic methodologies (explained in Sect. 4.2) to connect a group's Twitter account with its blog site and Facebook profile.
- To have a comprehensive understanding of a complex phenomenon (such as the ones discussed in this book), knowledge from multiple disciplines is needed. Hence, there is a need to encourage interdisciplinary courses and programs to equip researchers to be able to solve complex problems.

References

1. A. Pavlishak, *Water Supply Problem in Crimea to Cost $247—417 million - Kremlin Aide.* Available: http://tass.com/russia/729854
2. R. News, *Ukraine Builds Dam Cutting Off Crimea Water Supply.* Available: https://www.rt.com/news/158028-ukraine-water-supply-crimea/
3. N. Macfarguhar, *Aid Elusive, Crimea Farms Face Hurdles.* Available: https://www.nytimes.com/2014/07/08/world/europe/aid-elusive-crimea-farms-face-hurdles.html
4. S. Jerome, *Ukraine-Russia Conflict Results in "Water War".* Available: http://www.wateronline.com/doc/ukraine-russia-conflict-results-in-water-war-0001
5. S. Kumar, G. Barbier, M.A. Abbasi, H. Liu, Tweettracker: An analysis tool for humanitarian and disaster relief, in *ICWSM.* Available: http://bit.ly/2nx4gbc

6. M.A. Smith, B. Shneiderman, N. MilicFrayling, E. Mendes Rodrigues, V. Barash, C. Dunne, T. Capone, A. Perer, E. Gleave, Analyzing (social media) networks with NodeXL, in *Proceedings of the Fourth International Conference on Communities and Technologies* (ACM, New York, 2009), pp. 255–264. Available: http://dl.acm.org/citation.cfm?id=1556497

7. S. Al-Khateeb, N. Agarwal, Understanding strategic information manoeuvres in network media to advance cyber operations: a case study analysing pro-Russian separatists' cyber information operations in Crimean water crisis. J. Baltic Secur. **2**(1), 6–27 (2016)

8. N. Otan, *Trident Juncture 2015*. Available: https://jfcbs.nato.int/trident-juncture

9. L. Alexander, *Open-Source Information Reveals Pro-Kremlin Web Campaign*. Available: https://globalvoices.org/2015/07/13/open-source-information-reveals-pro-kremlin-web-campaign/

10. M. Bazzell, *Open Source Intelligence Techniques: Resources for Searching and Analyzing Online Information*, 4th edn. (CCI, Charleston, 2014). Available: https://inteltechniques.com/book1.html

11. B. Etling, J. Kelly, R. Faris, J. Palfrey, *Mapping the Arabic Blogosphere: Politics, Culture, and Dissent*, vol. 6. Available: http://www.ikhwanweb.com/uploads/lib/HNFNAB99APYNXAK.pdf

12. S. Al-khateeb, M.N. Hussain, N. Agarwal, Social cyber forensics approach to study twitter's and blogs' influence on propaganda campaigns, in *International Conference on Social Computing, Behavioral-Cultural Modeling and Prediction and Behavior Representation in Modeling and Simulation* (Springer, Berlin, 2017), pp. 108–113

13. S. Al-khateeb, M.N. Hussain, N. Agarwal Leveraging social network analysis and cyber forensics approaches to study cyber propaganda Campaigns, in *Social Networks and Surveillance for Society*, 1st edn. Lecture Notes in Social Networks (Springer, Cham, 2018) no. 2190–5428, p. 86. Available: https://www.springer.com/us/book/9783319782553

14. D.M.A. DoD News, *Operation Atlantic Resolve Exercises Begin in Eastern Europe*. Available: http://www.defense.gov/news/newsarticle.aspx?id=128441

15. R. T. *'Tanks? No Thanks!': Czechs Unhappy about US Military Convoy Crossing Country*. Available: http://www.rt.com/news/243073-czech-protest-us-tanks/

16. D. Sindelar, *U.S. Convoy: In Czech Republic, Real-Life Supporters Outnumber Virtual Opponents*. Available: http://www.rferl.org/content/us-convoy-czech-republic-supporters-virtual-opponents/26928346.html

17. Sputnik, *Czechs Plan Multiple Protests of US Army's Operation Dragoon Ride*. Available: http://sputniknews.com/europe/20150328/1020135278.html

18. S. Ghosh, B. Viswanath, F. Kooti, N.K. Sharma, G. Korlam, F. Benevenuto, N. Ganguly, K.P. Gummadi, Understanding and combating link farming in the twitter social network, in *Proceedings of the 21st International Conference on World Wide Web* (ACM, New York, 2012) pp. 61–70. Available: http://dl.acm.org/citation.cfm?id=2187846

19. F. Sen, R.T. Wigand, N. Agarwal, M. Mete, R. Kasprzyk, Focal structure analysis in large biological networks, in *IPCBEE*, ser. 1, vol. 70 (IACSIT, Singapore, 2014). Available: http://www.ipcbee.com/vol70/001-ICEEB2014-E0002.pdf

20. N. Agarwal, S. Al-khateeb, R. Galeano, R. Goolsby, Examining the use of botnets and their evolution in propaganda dissemination. J. Def. Strateg. Commun. **2**(2), 87–112 (2017). Available: https://www.stratcomcoe.org/nitin-agarwal-etal-examining-use-botnets-and-their-evolution-propaganda-dissemination

21. S.A. Tatham. *Strategic communication: a primer.* Defence Academy of the United Kingdom, Advanced Research and Assessment Group (ARAG) (Dec 30, 2008). Available: https://smallwarsjournal.com/documents/stratcommprimer.pdf

22. C. Staff, *ISIS Video Appears to Show Beheadings of Egyptian Coptic Christians in Libya*. Available: http://www.cnn.com/2015/02/15/middleeast/isis-video-beheadings-christians/

23. T.n. editorial, *ISIL executes an Israeli Arab after accusing him of been an Israeli Spy*. Available: http://www.tv7israelnews.com/isil-executes-an-israeli-arab-after-accusing-him-of-been-an-israeli-spy/

24. K. Shaheen, *ISIS Video Purports to Show Massacre of Two Groups of Ethiopian Christians*. Available: https://www.theguardian.com/world/2015/apr/19/isis-video-purports-to-show-massacre-of-two-groups-of-ethiopian-christians
25. S. Al-khateeb, N. Agarwal, *Examining Botnet Behaviors for Propaganda Dissemination: A Case Study of ISIL's Beheading Videos-Based Propaganda* (IEEE, Atlantic, 2015), pp. 51–57.
26. N. Abokhodair, D. Yoo, D.W. McDonald, Dissecting a social botnet: Growth, content and influence in twitter, in *Proceedings of the 18th ACM Conference on Computer Supported Cooperative Work & Social Computing* (ACM, New York, 2015), pp. 839–851. Available: http://dl.acm.org/citation.cfm?id=2675208

Glossary

Deviant Cyber Flash Mobs (DCFM) are defined as the cyber manifestation of flash mobs (FM), are known to be coordinated via social media, telecommunication devices, or emails, and have a harmful effect on one or many entities such as government(s), organization(s), society(ies), and country(ies). These DCFMs can affect the physical space, cyberspace, or both, i.e., the cybernetic space.

Focal Structure Analysis (FSA) is an advanced method/measure to discover a set of influential nodes or vertices (small community) in a large network or graph. This set of nodes do not have to be strongly connected and may not be the most influential on their own but by acting together they form a compelling power.

Online Deviant Groups (ODGs) are groups of individuals connected online, usually using social media platforms or the dark web and have interest in conducting deviant acts or deviant events that can cause significant danger to the public in general. These ODGs could include state and non-state actors, for example, Daesh, anti-NATO propagandist, Deviant Hackers Networks, and Internet Trolls among others.

Social Cyber Forensics (SCF) is a branch of cyber forensics (CF) and it's the process of investigating the relationships among "entities" (an information actor which can be a single individual, a group, an organization, etc.) and uncovers the hidden relations among them in social media space by extracting/collecting metadata associated with their social media accounts, e.g., affiliations of the user, geo-location, IP address, and WTC.

Web Tracker Code (WTC) is an online analytics tool that allows a website owner to gather some statistics about their website visitors such as their browser, operating system, and the country they are from, along with other metadata. These web trackers have an ID number that is usually embedded in the website HTML code. For example, Google provides users with a capability to track their website activities using Google Analytics service.

Index

Printed in the United States
By Bookmasters